IMAGES
of Aviation

BRITTEN NORMAN

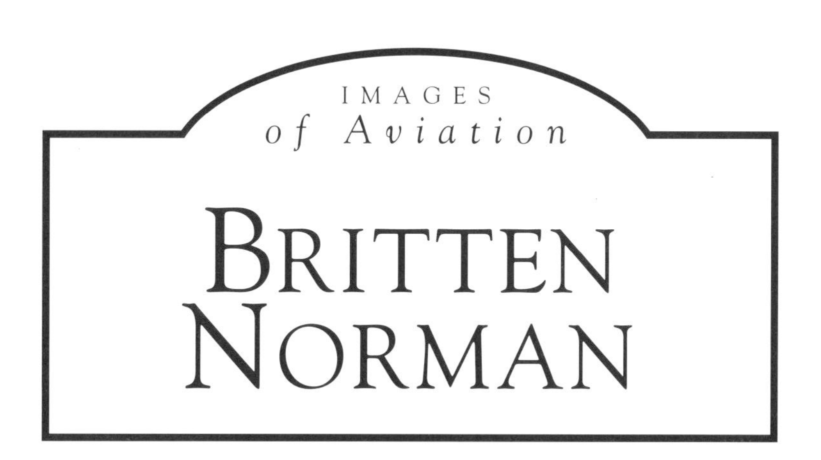

Compiled by
George Marsh

First published 2000

Tempus Publishing Limited
The Mill, Brimscombe Port,
Stroud, Gloucestershire, GL5 2QG

ISBN 0 7524 1729 0

Typesetting and origination by
Tempus Publishing Limited
Printed in Great Britain by
Midway Clark Printing, Wiltshire

Contents

Acknowledgements

My sincere thanks for racking their brains and memories, for checking the facts and in some cases for lending pictures go to Desmond Norman, Jim McMahon, Peter Gatrell, Sheila Dewart, Dave Williams, Jack Griffen, Ian Wilson, Bob Wilson, Guy Palmer, Peter Ward, Patrick Hallahan, Geoff Toms, Robin Maconachy of Bembridge Heritage Centre, plus Andrew Clancy and Allan Wright (BN Historians).

Some extremely valuable pictures from the early years were located and lent by Jim McMahon, Peter Gatrell, Dave Williams, Geoff Toms, the Bembridge Heritage Centre and the Britten-Norman Technical Department. For the subsequent years I have drawn on the extensive resources of the Britten-Norman archive under the care of Sheila Dewart, with contributions also from BN Historians.

Introduction

When two young apprentice aircraft engineers, John Britten and Desmond Norman, first met in 1947 at the highly regarded de Havilland Aeronautical Technical School at Hatfield, Hertfordshire, they little guessed how far their joint enthusiasm for aviation would take them.

A spare time partnership that formed when their training finished in 1949 was to evolve into an aircraft company whose main product would endear itself to users the world over for its rugged dependability, its ability to operate from crude airstrips and its remarkable flying qualities. Today, the ubiquitous Islander – along with the Defender and Trislander derivatives – remains a byeword for rugged dependability with dozens of operators worldwide.

For their first venture, the young entrepreneurs chose the de Havilland Tiger Moth as their vehicle to break into aerial agriculture. The agricultural Tiger Moth they evolved with another collaborator, Australian crop duster pilot Jim McMahon, offered the undemanding, operate-anywhere, easy-use, short take off and landing (STOL) and economical qualities that agricultural work required. Converting these slow-flying, ex-Ministry birds involved overhauling aircraft that had been picked up at auction for around £50 to £70 and adding to the front cockpit a McMahon hopper to hold the chemical spray. To start with, the partners worked in a shed at the bottom of John Britten's garden at Bembridge, Isle of Wight, but operations soon moved to roomier premises, the Unity Hall at the Commodore, cinema in Ryde.

In 1954, with demand for converted Tiger Moths in decline, the partners decided that inefficient crop spraying equipment then in current use could be improved. They developed a rotary atomizer which, under the name Micronair, subsequently became a commercial success, selling in thousands worldwide. In 1955 Britten and Norman, along with Jim McMahon and a fourth partner, Frank Mann, a West Country fruit merchant with aviation connections, formed two companies. One, Britten Norman Ltd, would concentrate on designing and building aircraft while the other, Crop Culture (Aerial) Ltd, would be an aerial spraying specialist. The four partners initially had equal shares in both. Beginning with two Tiger Moths in the Sudan, Crop Culture (Aerial) went on to operate more than ninety fixed and rotary wing aircraft, fitted with Micronair equipment, throughout the world. Britten and Norman eventually took over full control of the aircraft company, leaving the other partners to concentrate on the lucrative crop spraying activities.

While spraying plantations for banana shippers Fyffes and Elders, Britten and Norman started another venture combining their design, engineering and business skills. To improve the movement of bananas from plantations to port of shipment they developed a hovercraft. Cushioncraft 1, launched in 1959, was the first of a series which, during the 1960s and early 1970s progressed to the CC7. Although the hovercraft were never to prove a commercial bonanza, valuable experience was gained.

The first aircraft

The earliest Britten-Norman aircraft was the BN1F, an ultra-light monoplane which first flew from Bembridge airfield in May 1951. It was not successful and in 1953 the prototype was retired to a boathouse at Bembridge Harbour while its Lycoming engine went into a wind tunnel made for testing Micronair atomizers. Later restored, G-ALZE is now on display at the R.J. Mitchell Hall of Aviation in Southampton.

Islander

Fortunately, the partners tried again, first conceiving their second aircraft as a replacement for de Havilland's venerable Dragon Rapide commercial twin-engine biplane. A potent influence was their experience with a BN associate company, Cameroons Air Transport, in operating light American aircraft in Africa. These could not carry more than four or five people and performed poorly in a hot climate. A search for a light utility twin having short take-off and landing characteristics and costing less than £25,000 proved fruitless, leading John Britten to conclude, 'We must design our own plane.'

Against the performance-dominated design trends of the time, Britten and Norman decided to keep their aircraft as simple and affordable as possible. A high wing of regular shape would not only ensure excellent pilot and passenger visibility but could be constructed in one piece. A near-rectangular box fuselage with simple nose and after sections could, likewise, be built cheaply. Basic, carburetted, six-cylinder piston engines widely used in other aircraft would be familiar to mechanics anywhere in the world. A large single fin and elevator would be straightforward and easy to fabricate. A fixed tricycle undercarriage offered strength with simplicity. Alloy construction, controls and instrumentation would all be conventional.

A large wing area gave low structural loadings and exceptional fatigue life. With its characteristic single slotted flaps and ailerons, the BN2 would be able to land at 65 knots and fly at less than 50 knots without stalling. It would seat ten, or carry fewer with freight.

Full design began in November 1963 and the go-ahead for a prototype was given three months later. This aircraft, G-ATCT, was completed within nine months and a maiden flight, with Britten and Norman at the controls, followed on 13 June 1965. Afterwards Desmond Norman declared: 'Absolutely delightful. She flies like a fully developed aircraft that has been in service for a couple of years.'

Further testing, however, revealed a need for greater power, so the two 210hp Continental IO-360-B piston engines first fitted were replaced by 260hp Lycoming O-540E units. At the same time, the wing span was extended by four feet. The resulting modified prototype, which first flew in December 1965, was christened the 'Islander' the following summer.

Corporately, Britten-Norman has had a roller-coaster ride. After funding the first eighteen months of development itself with the help of Crop Culture proceeds, the company needed outside finance. City investors could not be tempted but the Ministry of Aviation was persuaded to provide half the launch costs, on condition that the government would receive a levy on sales. (The investment proved sound as the Islander would become the only British aircraft ever to repay government launch aid in full, with interest.)

When it looked as though the second aircraft and first production prototype would not be ready in time for the 1966 Farnborough Air Show, Britten-Norman's 300 employees at Bembridge sacrificed their annual holiday to make it happen. Shortly after this they moved to a new 56,000sq.ft factory which replaced two workshops where fuselages and wings had previously been built separately.

Tragedy struck in November 1966. Test pilot Peter Hillwood died when G-ATCT crashed in Holland after encountering severe icing. The design was, however, exonerated and a build up in production continued.

Launch customers GlosAir (who formed Alderney-based Aurigny Air Services) and Scottish operator Loganair accepted their first Islanders after the type was awarded its Certificate of Airworthiness in August 1967. Both operators quickly put their BN2s into passenger service and have continued making money with Britten-Norman products ever since. Orders for the light utility twin, first priced at £17,300, came flooding in. One, from the distributor for the Americas, Jonas Aircraft of New York, was for thirty aircraft.

Record breaker

Many doubted Britten-Norman's early prediction that 1,000 aircraft would be sold within ten years. By the time the 100th aircraft – one of ten for Papua New Guinea – was delivered in 1969, the testimonials of fifty-six operators in twenty-seven countries were making the sceptics think again. In 1970 BN gained a Queens Award to Industry for Export Achievement. It took just five more years for the Islander to break the British sales record for a multi-engine commercial aircraft, previously held by the DH Dove with 548 units.

Admittedly, it took fifteen years rather than ten to sell 1,000, but today over 1,200 Islander and derivatives have been sold, making the marque Britain's most successful ever in terms of aircraft numbers. Remarkable also is the fact that some 800 of these are still in service around the world, such is the Islander's durability.

Islanders have been marketed direct, traded for goods, sold under a range of finance packages and funded by World Bank loans. Customers have taken over demonstrators and aircraft have even been given away – a BN2B for Mali was paid for by Live Aid following a concert by pop star Bob Geldof.

Turbulence

Troubled times brought repeated location and ownership changes, fortunately without halting the Islander's success.

As orders passed the 200 mark, Britten Norman started to subcontract first wings and then entire aircraft kits to the British Hovercraft Corporation at Cowes, leaving Bembridge responsible for final assembly. BHC went on to produce some 370 aircraft but, following a purchase by Romania of British BAC 1-11 airliners, primary production was transferred to that country under a national offset deal. The first Romanian-built Islander flew in 1969. BN aircraft are still built at Bucharest to this day.

In the early 1970s a mismatch between sales and production led to cash flow difficulties prompting backers to withdraw credit. This forced the company into receivership, though trading returned to profitability during the ensuing year. Fairey SA of Belgium acquired BN as a going concern, moving primary production to its factory at Gosselies. Fairey Britten-Norman went on to build aircraft at a record rate of almost three a week, Fairey's military background proving helpful in securing such orders as a £1m sale of eight Defender (military Islander derivative) aircraft to Oman.in 1974. The first of these was BN's 500th aircraft delivered. Under a contract for 100 Islanders concluded in the same year, the Philippines also became an important manufacturing centre.

John Britten and Desmond Norman left the company in 1976 to pursue other interests. Three years later the aviation world was saddened when John Britten died, aged only forty-nine. He was buried at Bembridge Church on his beloved Isle of Wight.

When Fairey themselves were forced into receivership in 1978, the Britten-Norman baton was taken up by Swiss aircraft company Pilatus. Primary production was returned to Romania, with Bembridge becoming the customization and finishing centre and Pilatus Britten-Norman headquarters. The Swiss agenda was to raise the added value of BN products, with Islander upgrades, special role variants and increased emphasis on military derivatives incorporating advanced surveillance equipment. Worldwide financial stringency and the end of the Cold War made this approach less attractive and in 1998 ownership of Britten-Norman passed from Pilatus to investment group Biofarm Inc.

Two years later the corporate future was in doubt as, in an uncanny repeat of history, receivers were once again called in follow cash flow difficulties. At time of writing, further change of ownership is in prospect.

Evolution

John Britten and Desmond Norman's deceptively simple airframe concept has proved endurable and adaptable to changing market needs.

Refining the early BN2 with increased payload, a larger baggage bay, an updated interior and other improvements resulted in the BN2A. Increasing the wing leading edge camber enhanced short-field performance.

A fuel-injected version of the Lycoming piston engine delivered 300hp rather than 260hp, improving performance in hot, high locations. But the biggest powering change came with the 320hp turboprop engines of the BN2T turbine Islander. The Allison 250-B17C-powered aircraft, which first flew in 1980, is faster, takes off and climbs more quickly, carries more and burns Avtur (kerosene) rather than the often more expensive Avgas (petrol).

Upgrading the standard BN2A to BN2B during the Pilatus era gave increased landing weight, a new instrument panel and enhanced passenger comfort through air conditioning and better sound-proofing. Britten-Norman has developed many options including three-blade propellers (quieter than the standard two-bladers), luxury VIP interiors, wing-tip tanks and special role fits.

The Islander's robust 'Land Rover'-like utility appealed to military and law enforcement users and the Defender variant was introduced in 1971. A strengthened airframe with underwing hard points make the Defender adaptable to target towing, parachute training, maritime surveillance, battlefield observation, artillery spotting, command control and communications, utility, VIP transport and many other roles. Maritime Defenders can carry anti-shipping missiles. Twin engines enable law enforcement Defenders to operate safely over the sea or urban areas.

The market desire to carry more passengers economically led Britten-Norman to lengthen the Islander's fuselage, extend the wing and add a third engine, mounted on the tail. The resulting eighteen-seat Trislander made its Farnborough Air Show debut in 1970. Aurigny became the first scheduled operator a year later. Though production ceased in 1983, around forty Trislanders remain in service throughout the world. Recently the company announced that production could resume, initially for China where operators regard Trislander as uniquely cost-effective.

Another airframe and wing extension, but using two turbine engines, led in 1998 to the appearance of the Defender 4000. Initially purchased by police forces in UK and Ireland, as well as commercial operator Sabah Air, this aircraft has attracted widespread law enforcement interest.

Although the Islander has set the global standard for flexible and economic short-sector service, often – appropriately – between islands, John Britten and Desmond Norman's brilliant creation has adapted well beyond this role. Its remarkable versatility, along with the highlights of its rise to become Europe's most successful light utility twin, are illustrated in the following pages.

One

The Early Years

The first aircraft to be designed and built by partners John Britten and Desmond Norman, *c.*1950, was this ultra-light, single-engine monoplane. The 'Finibee' (the name was a 'reverse' interpretation of the designation BN1F), G-ALZE, initially had a 36hp Aeronca Jap engine but later a 55hp Lycoming was introduced, along with increased wing span. It first flew in its more powerful form from Bembridge Airfield in 1951. However, the type was not successful and only one example was built. After languishing for years in a boathouse at Bembridge, this aircraft was refurbished and is currently on display in the R.J. Mitchell Hall of Aviation, Southampton.

Previous page: Desmond Norman (left) and John Britten at Bembridge with the aircraft that made them famous. Desmond Norman, born in August 1929, came from a Gloucestershire family, was educated at Eton and the de Havilland Aeronautical Technical School and was a keen pilot. He served for eight years with the Royal Air Force, at one point flying jets as a fighter pilot with the squadron his father, Sir Nigel Norman, had commanded before the Second World War. A natural marketer, he was for two years export assistant with the Society of British Aircraft Constructors and altogether flew more than fifty aircraft types. John Britten, born in May 1929, was educated at Dartmouth Naval College and the de Havilland Aeronautical Technical School, where he first met Norman. From a Bembridge family, he had a designer's mathematical mind and became an Associate Fellow of the Royal Aeronautical Society. By the mid-1960s he had flown some 600 hours in light aircraft. The BN2 Islander became Britain's most successful commercial passenger aircraft, with over 1,200 sold. More than 800 remain in service worldwide.

Britten-Norman's first employee, Peter Gatrell, is seen (centre) with John Britten (left) examining the engine installation on the Finibee in the original works, a hangar/workshop next to the Propeller Inn on the south side of the airfield. Gatrell also worked on Tiger Moth conversions, then hovercraft and was later to became a manager with Crop Culture (Aerial) Ltd.

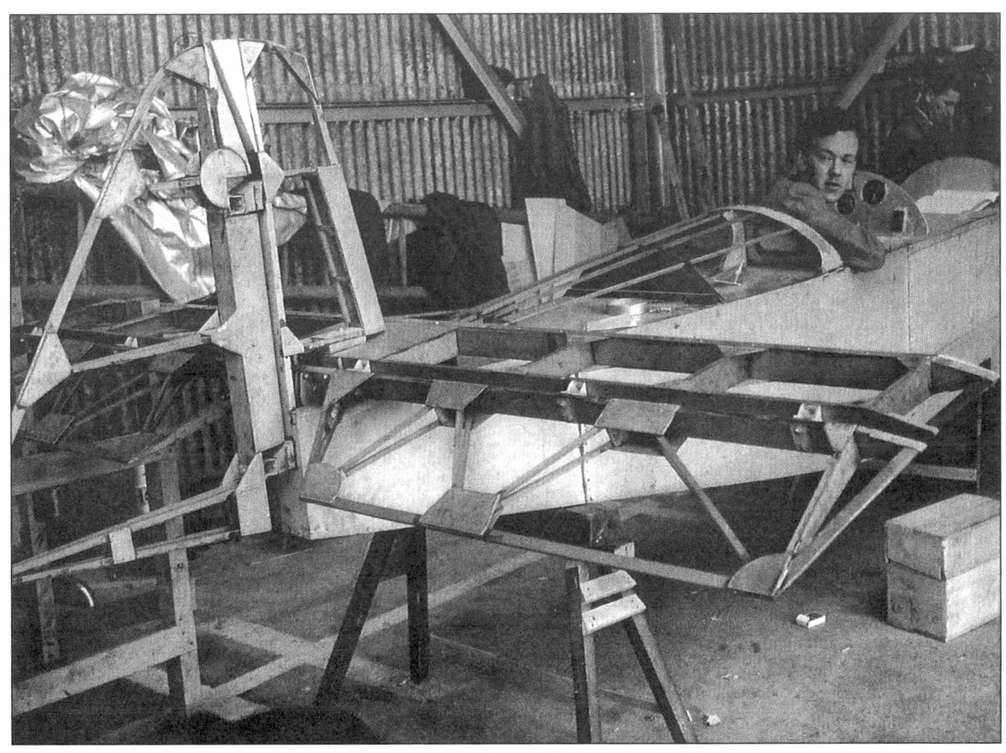

Here John Britten checks the elevator controls on the BN1F Finibee.

Test running the engine meant first taxiing the wingless Finibee to these fuel pumps at the flying club, on the other side of the Propeller Inn. A short while later, the partners took on the management of the airfield, including the inn and the flying club whose premises are seen in the background.

The partners carried on some of their Tiger Moth conversion business in a shed in the garden of John Britten's one-time home, 'St Denis', in Bembridge High Street. (The shed is still there, behind the house.) This was where they would complete McMahon hoppers and other agricultural equipment for installation in aircraft which had become surplus to Ministry requirements. Some of the Tiger Moths, clearly well used, had to be patched up with brown paper and Lascovic tape before they could be flown to the Island. Local reports suggest that the more disreputable examples were flown in at night! Many completed aircraft were exported to New Zealand for top dressing work, but this example is seen spraying potatoes near Stapleford in Essex, UK.

Converting surplus Tiger Moth biplanes for agricultural use provided Britten-Norman's first serious revenue. The de Havilland classic possessed the robust simplicity, low maintenance and excellent slow flying and short-field capabilities that the partners would later strive for in their own designs. The leather-helmeted pilot in the cockpit of this example, c.1954, is Jim McMahon. Note the Micronair airsteam-driven atomizer/spray equipment.

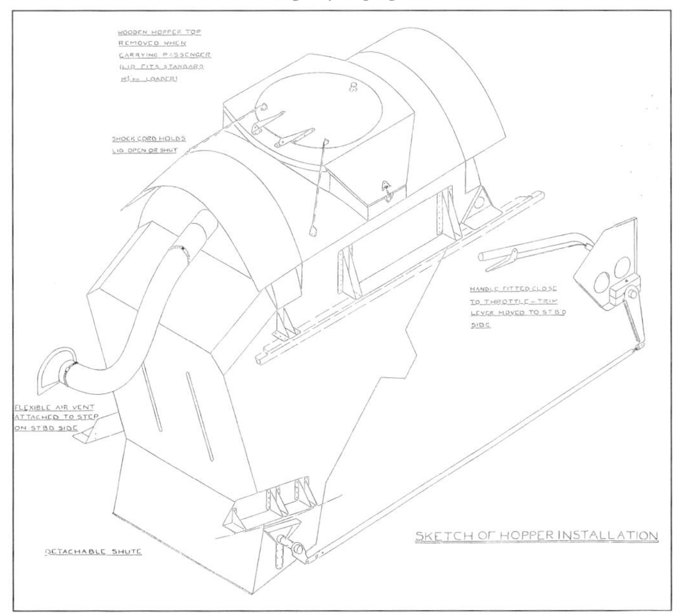

The McMahon hopper, shown in this sketch from an early Britten-Norman brochure, could hold 5cwt (as advertised) of top dressing fertilizer and was installed in the cockpit of a DH 82 Tiger Moth. The pilot used a control lever near the throttle to release fertilizer in a low-volume, controlled flow, usually at the rate of about 1cwt per acre. The system was approved by de Havilland and the Air Registration Board.

Eventually all conversion operations were based at Unity Hall, then beneath the Commodore cinema in Ryde, one of several cinemas owned at that time by the Britten family. John Britten knew the premises well having at one time worked as a projectionist there. Local people became familiar with the sight of wingless Tiger Moth fuselages being towed backwards by Mrs Britten in her Ford Prefect, as each was transferred to the airfield for final assembly and flight. Here a Tiger Moth fuselage is manoeuvred into Unity Hall from the street.

The Micronair compact rotary atomizer revolutionized aerial crop spraying and was the foundation for the success of Crop Culture (Aerial) Ltd, sister company to Britten-Norman during the 1950s. John Britten and Des Norman helped develop the efficient, fine droplet spray generator and were partners in Crop Culture, which began with two Tiger Moths in the Sudan but became extremely successful and went on to operate over ninety aircraft worldwide. Later, Micronair equipment was fitted to a range of different aircraft types including the Islander, as seen in this example where the installation includes a large underwing tank for the spray chemical.

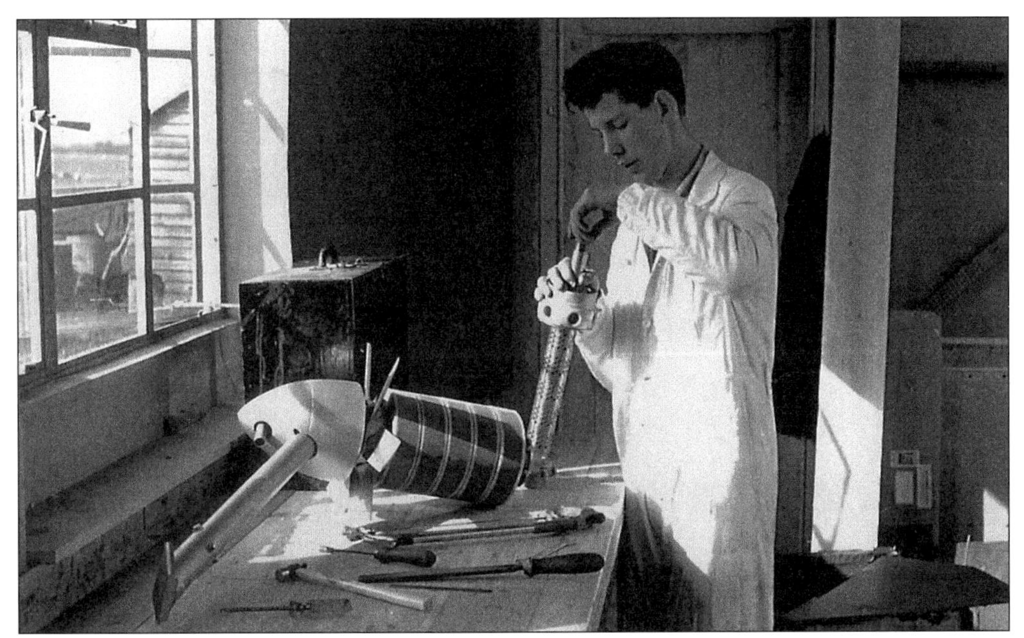

The Propeller Inn has played an essential part in Britten-Norman's history. Here an apprentice assembles an AU2000 atomizer in what is now the Propeller Inn bar.

Britten-Norman took part in the hovercraft fever of the 1960s. Its fully amphibious Cushioncraft series, first conceived for transporting bananas from plantations, was quieter than other hovercraft of the period thanks to a patented, centrifugal, fan-based lift and propulsion system designed and produced by associate company Cushioncraft Ltd. The first in the series, CC1, was a small craft powered by a 160hp Coventry Climax sports racing engine. This was only the second working hovercraft ever to lift off. Here it is seen hovering, with Desmond Norman at the controls and John Britten and Peter Gatrell alongside.

CC2, pictured here, was a similar but larger and improved machine. Note the circular shape and the absence of any skirt, which was a later hovercraft innovation. Britten and Norman are at the right of the picture.

Dave Williams, the second person to be employed at Britten-Norman, is seen here working on CC2.

Some of the earliest employees of Britten Norman Limited are pictured here in the original building next to the Propeller Inn, where fuselages were later built for the first Islanders. Cushioncraft work took place there too – on the floor can be seen the jig for the fan on CC1. Pictured in the late 1950s are, from left to right, Jim Young, George Smith, Trevor Young, Dave Williams, Stan Wright, Jack Morrell, Pete Sothcott, Malcolm Barclay, 'Wac' Walker, John Locke and Pete Preston.

By the time the later Cushioncraft were 'flying', the BN2 Islander was well into its stride too. Here an Islander paces a ten-seat CC7 off the Isle of Wight, c.1969.

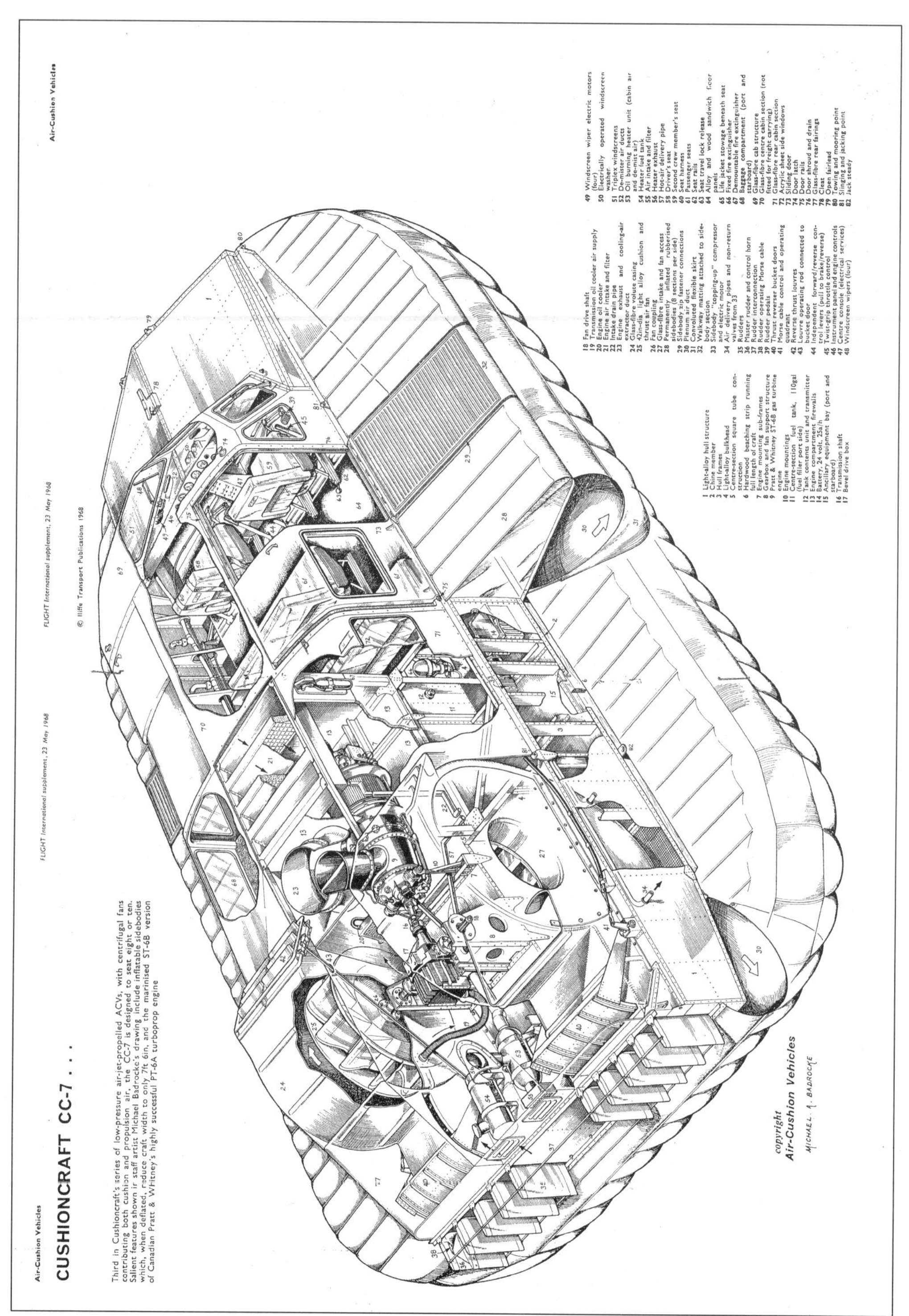

This cutaway drawing of the CC7 shows that its concept was advanced for the era, the gas-turbine-driven centrifugal fans providing both lift and propulsion. British Hovercraft Corporation at Cowes later purchased Cushioncraft, but failed to develop its products further.

The partners' second aircraft venture was altogether more ambitious than the Finibee and their light utility twin would become a world best seller. Here the first prototype, G-ATCT, is being prepared for its maiden flight, from Bembridge. The aircraft took off on 13 June 1965 with Desmond Normand and John Britten at the controls and made a successful seventy-minute flight. Four days later, under special certificate of airworthiness clearance, it flew to Paris to appear at the Air Salon.

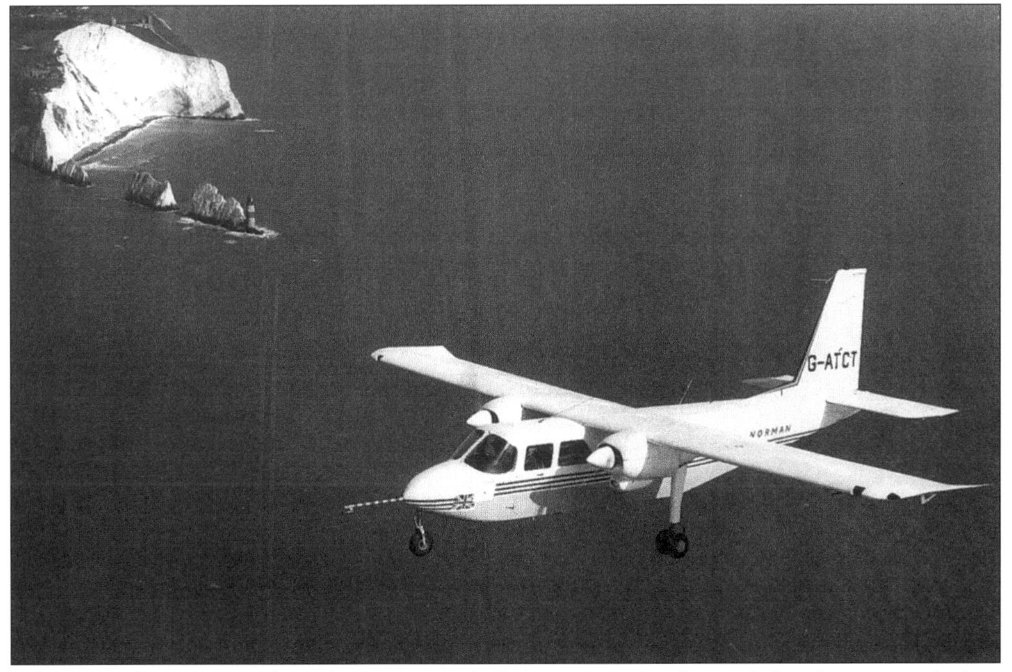

Subsequently re-engined, with Lycoming 0-540 engines in place of the Continentals first fitted, and with its wing span extended by 4ft, the prototype first flew in its modified form on 17 December. Unfortunately G-ACTC, seen here off the Needles, Isle of Wight, during the mid-1960s, was lost in an accident in the Netherlands in November 1966. Test pilot Peter Hillwood was killed in the crash, caused by severe icing.

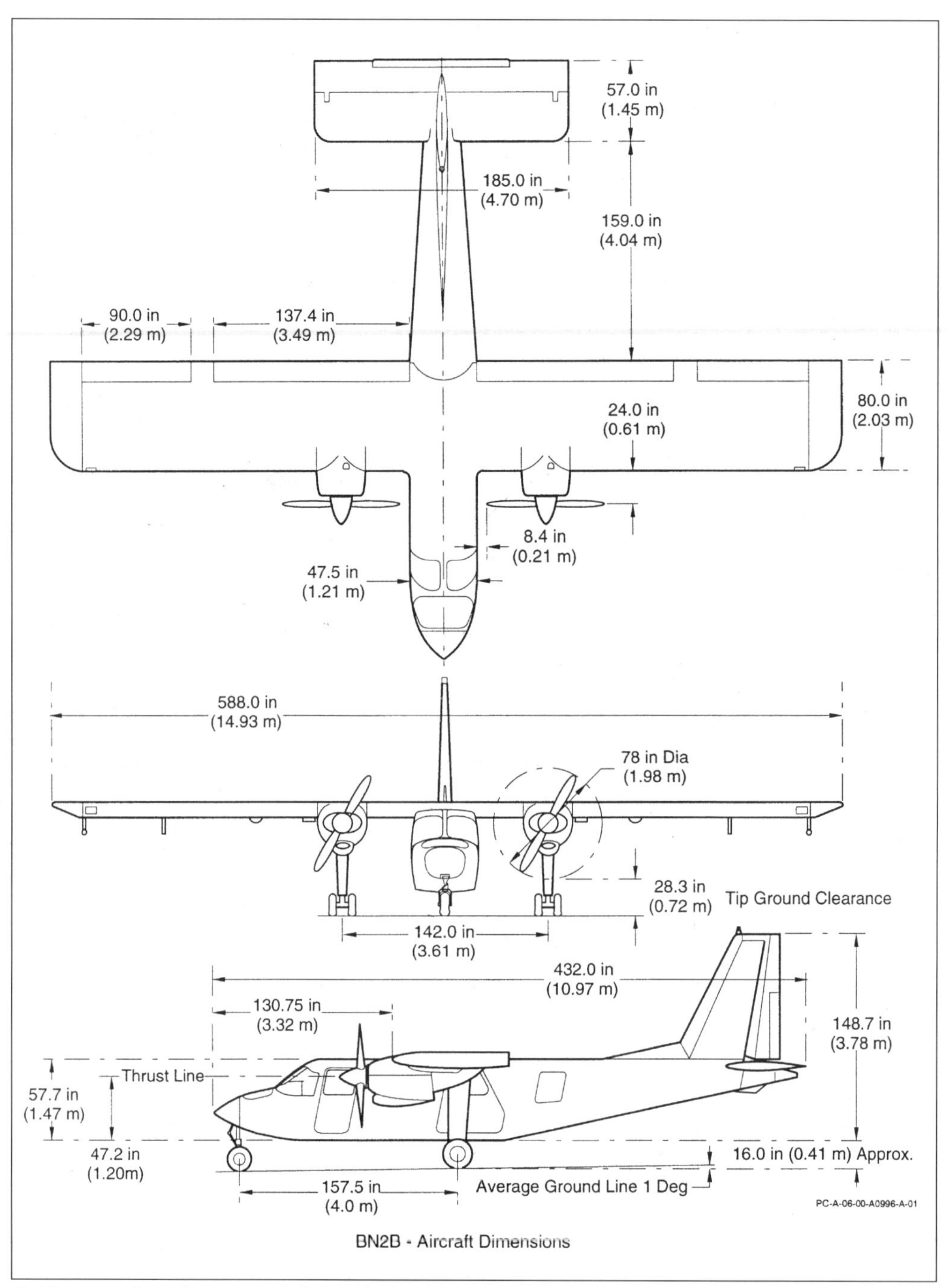

As these outline dimensions show, the BN2 was conceived as a compact light twin that could serve as a utility aircraft and mini airliner.

The Islander airframe has always been highly regarded as being tough and practical. Here Desmond Norman and chief designer Denis Berryman (in front), with John Britten and Ernie Perkins behind them, try out a wooden mock-up for size in the Propeller Inn works. Note the largely rectangular basic shape of the fuselage, making for easy production and low cost in conventional alloy construction.

Later, when the success of the new type became clear, operations moved to a new, purpose-built factory on the east side of the airfield where volume production could take place. Our photograph shows a typical scene once production had built up in this facility.

An Islander is seen in its 'home territory' over Bembridge airfield, with Bembridge village and harbour in the background.

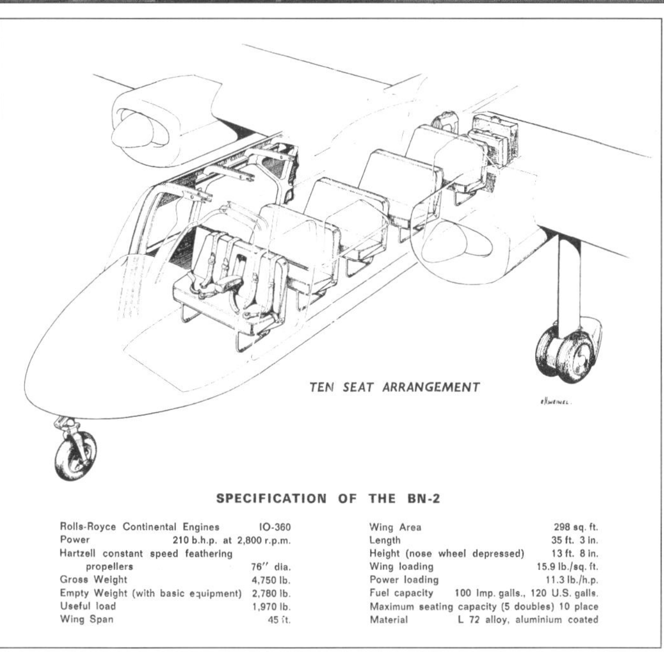

SPECIFICATION OF THE BN-2

Rolls-Royce Continental Engines	IO-360	Wing Area	298 sq. ft.
Power	210 b.h.p. at 2,800 r.p.m.	Length	35 ft. 3 in.
Hartzell constant speed feathering propellers	76″ dia.	Height (nose wheel depressed)	13 ft. 8 in.
Gross Weight	4,750 lb.	Wing loading	15.9 lb./sq. ft.
Empty Weight (with basic equipment)	2,780 lb.	Power loading	11.3 lb./h.p.
Useful load	1,970 lb.	Fuel capacity	100 Imp. galls., 120 U.S. galls.
Wing Span	45 ft.	Maximum seating capacity (5 doubles)	10 place
		Material	L 72 alloy, aluminium coated

An excerpt from an early brochure showing the basic ten-seat arrangement and specifications for the BN2 Islander. Note the baggage bay behind the passenger seating.

4 FLIGHT International, 24 August 1967

COMMERCIAL — PROFITABLE — RUGGED
— STOL — SIMPLE — SHORT HAUL — COMMERCIAL
PROFITABLE — RUGGED
— STOL — SIMPLE

$62,000 / 10 SEATS = $6,200 PER SEAT

FOR OPERATING DATA

IN U.K. WRITE TO:—

BRITTEN-NORMAN LTD.
BEMBRIDGE AIRPORT,
ISLE-of-WIGHT, ENGLAND.

OR TELEPHONE:— BEMBRIDGE 2511
OR CABLE:— BRITNOR, BEMBRIDGE

IN U.S.A. WRITE TO:—

JONAS AIRCRAFT CO., INC.,
120 WALL STREET,
NEW YORK.

OR TELEPHONE:— 212 WHITEHALL 4-4517
OR CABLE:— JONASNEL, NEW YORK.
OR TELEX:— TWX 212-571-0857

This 1967 advertisement for the Islander emphasizes the low capital costs of the BN2A in the universal aviation currency, the dollar.

Two
Trials and Tribulations

Production in volume has often proved a difficult challenge for Britten Norman, involving several changes of location and ownership. A number of aircraft were built at Bembridge, particularly in the early days, but many more have been manufactured at Cowes (British Hovercraft Corporation), in Romania (Intreprendirea de Reparat Material Aeronautic [IRMA] – later Romaero SA), in Belgium (Fairey SA) and in the Philippines (National Aero Manufacturing Corporation). Bembridge, though, has always been responsible for finishing to customer requirements and certification.

This Trislander is one of several whose manufacture took place at Bembridge.

Initially the BN2 (christened 'Islander' in 1966) was a private venture but, after persuasion, the government stepped in to help fund production in return for a levy on aircraft sold. One result was this 56,000sq.ft factory, completed in 1966, which replaced two workshops where wings and fuselages had previously been produced separately. One of these was the Propeller Inn works on the airfield and the other was at a Bembridge Harbour boatyard.

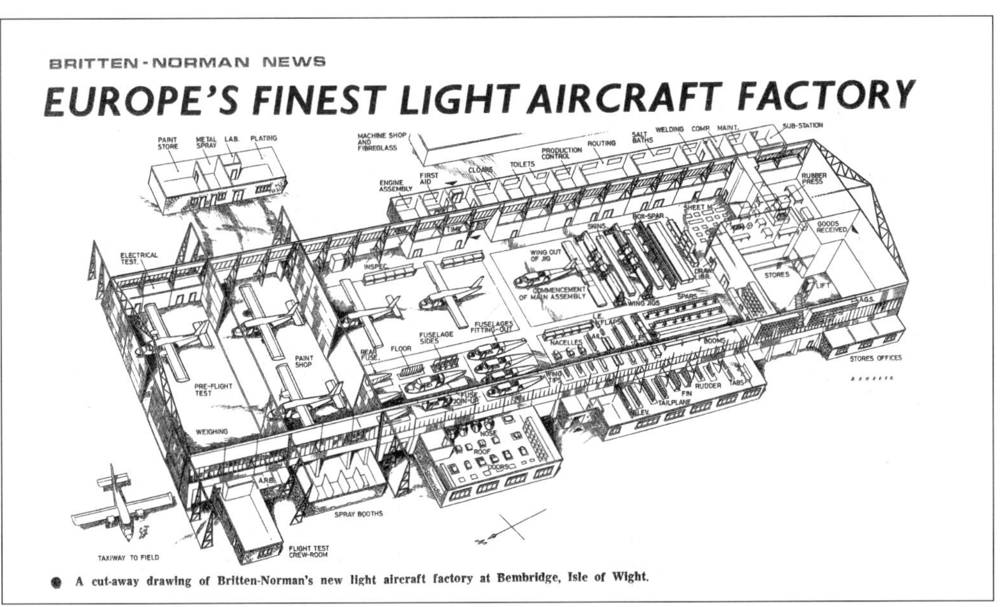
BRITTEN-NORMAN NEWS

EUROPE'S FINEST LIGHT AIRCRAFT FACTORY

● A cut-away drawing of Britten-Norman's new light aircraft factory at Bembridge, Isle of Wight.

This is how the house newspaper *Britten Norman News* described and illustrated the new £250,000 factory in 1967. It was partly 'self-built' since direct labour from BN and associates were responsible for the fuselage shop. The first aircraft to leave the facility was a BN2 for GlosAir, who took delivery in late summer 1967.

Bembridge Airport has existed since the 1930s, for many years as just a grass strip. John Britten and Desmond Norman began flying their Tiger Moths from it in the early 1950s and at the start of the Islander era began to manage it for the local farming interests who owned it. The 2,950ft concrete runway was added in 1980 and a new terminal building was erected in 1987.

Today the airport has air traffic control, customs, refreshments and taxi facilities, with Britten-Norman encouraging its use as a general aviation facility. Bembridge has become a popular venue for private flying events and once a year hosts the relaunched Schneider Trophy air race. Here, BN marketing communications manager Sheila Dewart is seen with Peter Wilson, then marketing manager, during an event at the airfield.

John Britten and Desmond Norman (standing) can afford to smile, as the company's first traumatic brush with receivership comes to an end. Receiver Maurice (Monty) Eckman of Price Waterhouse (sitting, right), who kept Britten-Norman trading profitably for a year after its insolvency in 1971, is signing the company over to the Fairey Group represented by managing director Bob Holder. Fairey, who had not produced any complete aircraft since the 1950s, moved production and flight testing to their facility at Gosselies, Belgium. Under the agreement Britten and Norman left the company in 1976, having helped to ensure its successful integration with Fairey Aviation. Sadly, John Britten was to die less than three years later after a heart attack.

Above and below: Freak 'twisters' in December 1978 wrought havoc among dozens of 'green' engineless airframes awaiting sale and fit-out at Bembridge. Ninety mile an hour winds broke lashings and overturned a number of aircraft, wrecking those and several others. Thanks to their structural simplicity, most of the Islanders were able to be repaired, though four had to be written off.

By the mid-1970s, the Bembridge factory was a fitting out and customization centre for aircraft built elsewhere. The aircraft in the foreground are BN 2A-21 Defenders destined for 550 Squadron of the Indian Navy at Cochin.

Britten-Norman's second experience of receivership occurred when Fairey themselves came to grief in 1977. In part they were caught out when an unexpected decline in customer demand coincided with a period of peak production. Again, Britten-Norman was sold on as a going concern, this time to Pilatus Aircraft of Switzerland. Here Mr. D.C. Klöckner (left), managing director of Pilatus, who assumed the same role at Pilatus Britten-Norman, discusses technicalities with Dr Bührle of the Oerlikon Bührle Group (owners of Pilatus) and BN chief designer Dennis Berryman. They are in the BN drawing office.

During the Pilatus era, Britten-Norman upgraded the Islander from the 2A model, which had seen few changes for some years, to the BN2B. This had an improved interior and other 'customer-friendly' features. The first BN2B is seen here at the Needles, Isle of Wight, having made its maiden flight in August 1978.

PBN moved primary production back to Bucharest. The Romanians had first started producing Islanders in 1968 as part of a contract at that time to build 215 examples. Under the renewed relationship, IRMA – later Romaero – was initially asked to build 100 of the new BN2B model. By 1993, a quarter of a century after it began, Romanian production had averaged some twenty per year, with almost 500 examples built. Today Romaero still produces green aircraft which are flown to Bembridge for finishing. The 500th BN2 to be built in Romania was this 2T (turbine) delivered green to Bembridge, where it is seen with Britten-Norman pilots Iain Young and John Ayers. The latter was chief test pilot at the time and Iain became chief test pilot later.

Sir Mark Norman, older brother of Desmond, became chief executive of the Islander production programme in Romania in 1968, having previously been managing director of the associate company Britten-Norman Sales Ltd in Chelsea, London. Here Sir Mark is seen with Alexander Onassis, then chairman of BN customer Olympic Airways.

Britten-Norman has always attracted its fair share of attention from dignitaries. Here Paddy Ashdown, British Liberal Democrat leader, shakes hands with Anthony Stansfeld, PBN managing director from 1992 to 1997, during a visit to Bembridge. With him is Dr Peter Brand, later to become the Liberal Member of Parliament for the Isle of Wight.

Jim Heath of Westinghouse, a BN partner at the time, greets a rising star in American politics, one Bill Clinton, against the backdrop of an advanced surveillance Islander, on display in the USA.

Anthony Stansfeld with the Duke of Kent at Bembridge, May 1997.

Earl Mountbatten is seen with BN's Lee Smith in 1976 on the occasion of a presentation to BN of a Queens Award for Industry. In the background is a Trislander destined for Trans Jamaican Airlines.

The Mayor of South Wight, Mr H. Howe, congratulates Warrant Officer Chris Sherfield (left) and Major George Bacon, subsequently of BN's marketing department, after their return with this British Army BN2T Defender from deployment in the Arabian Gulf during the war there in the early 1990s.

There have been royal visitors too. Here, during a visit to Bembridge in the early years by the Duke of Edinburgh, are also, from left to right, Jim Birnie (chief test pilot), Jim McMahon, Desmond Norman and Isle of Wight MP Mark Woodnutt.

Film director Richard Attenborough created a film set at Shepperton Studios when making his successful 'Cry Freedom', in which leading characters are involved in an escape using an Islander. He used a fuselage from the first Romanian-built Islander, which had been damaged in an accident during 1974. The fuselage was shipped to the UK with BN's assistance.

During 1971, John Britten and Desmond Norman briefly visited Ceylon during a visit to the Far East to assess the Islander's potential in the area. Here they are seen at an official function.

A decade later, the Cypriot government had become a customer and in 1982 it took delivery of the 1,000th Islander/Defender. G-MICV , piloted by chief test pilot Hugh Kendall with Peter Ward as observer, is seen here towing an aerial practice target prior to its flight to Cyprus.

At a short ceremony to mark the acceptance of the 1,000th Islander at Bembridge are, from left to right, Reg Caudle, five Cypriots representing customer interests, Peter Wilson, David Fear, Savvas Constantinedes (Greek agent), Andy O'Connell, Bob Wilson, Terry Hollins, Trevor Ward, Malcolm Gould and another BN employee (identity unknown).

All the trials and tribulations seemed worthwhile at the thirtieth anniversary, in June 1995, of the Islander's maiden flight. The UK and overseas operators flying in for the occasion constituted one of the largest working Islander/Trislander gatherings ever seen at Bembridge. Desmond Norman was a special guest. Commenting on the occasion, managing director Anthony Stansfeld said, 'There aren't many aircraft types that have been flying for thirty years and look like remaining successful for another thirty.'

G-BDWV

Three
Broadening the Concept

Loganair found the Trislander a perfect complement to their existing Islanders.

So distinctive and popular are Aurigny's Trislanders that one of them, G-JOEY, has also become a cartoon character called Joey.

Previous page: Britten and Norman's simple, economical, 'no gimmicks' approach to aircraft design has proved both durable and adaptable. One of several variations on the theme is the BN2A MkIII Trislander, achieved by enlarging the Islander airframe (with a lengthened fuselage and extended wings) and adding a third, tail-mounted engine. Channel Islands-based Aurigny Air Services became the first scheduled operator in 1971 and is still using the type almost thirty years later.

It proved feasible to mount a third Lycoming piston engine on Trislander's fin structure, in the manner of many 'big jets' of the time, as this American example shows.

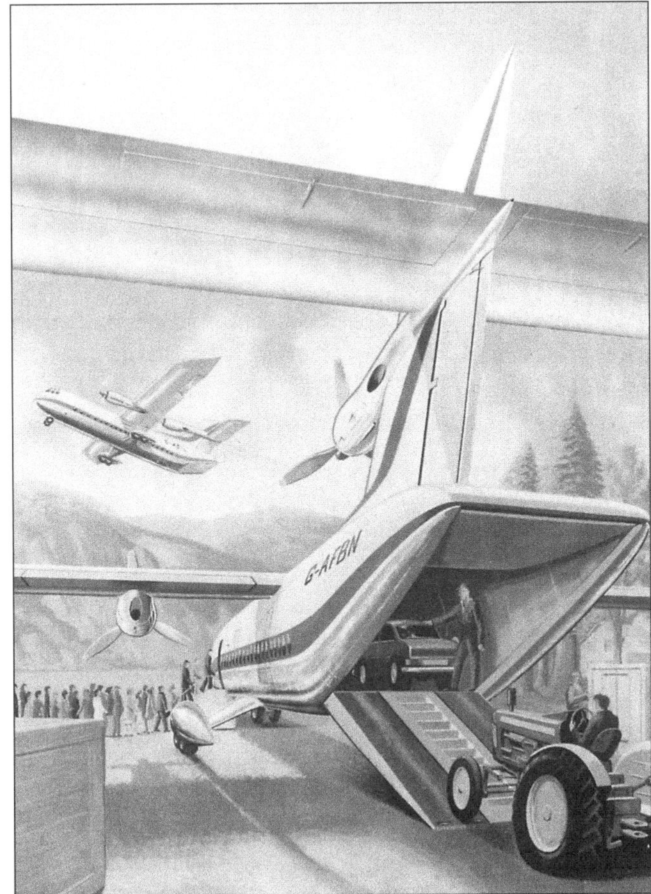

Greatly enlarging the three-engine, high-wing concept resulted in the BN4 'Mainlander', a transport aircraft first proposed as a Bristol Freighter replacement. Again, simplicity was the key with no pressurization, a fixed undercarriage, conventional structure and three Rolls Royce Dart turboprops. Mainlander would have carried up to 100 passengers or ten tons of freight/vehicles, with various mixed configurations possible. Despite interest from British Air Ferries and other operators, the necessary financial backing was unavailable in 1972 and the ambitious project never progressed beyond the feasibility study stage.

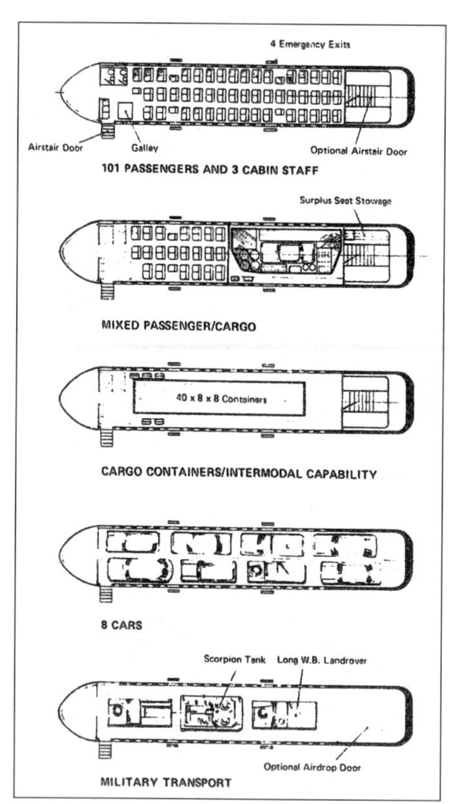

The versatility characteristic of Britten-Norman designs would have been shared by the Mainlander, as these alternative layouts show.

Islanders have been given progressively more power. Six-cylinder, carburetted Lycoming engines like this started off at 260hp. The classic piston engines are easily maintained by mechanics with basic skills anywhere in the world. They are petrol driven and highly reliable. Introducing 300hp Lycoming IO 540-K1B5 engines with fuel injection added a sparkle to the Islander's performance with a nine-knot faster cruising speed and a higher ceiling. For hot and high conditions, BN distributor Jonas developed the Rajay supercharger, raising the single-engine ceiling to 12,500ft.

A more fundamental powering change was the development of a turboprop variant, resulting in the BN2T Turbine Islander – visually distinguishable by its air inlets over the top of the engines rather than underneath. Turbine benefits include higher speed, faster climb and immunity to periodic shortages of Avgas fuel. The aircraft are also quieter. The example pictured became a prototype BN2A-41 Turbine Islander, being fitted with Lycoming LTP101 turbine engines in 1977. However, these engines proved too powerful and were not finally selected.

The engine eventually chosen was the 320hp Allison (later Rolls-Royce-Allison) 250-B17C. Here a turbine engine is seen on an airframe modified at Bembridge to receive it. Prototype G-BPBN first flew in August 1980 and received its UK certification in May 1981. The aircraft was delivered to Frank Jonas, BN's distributor for the Americas, by the end of that year.

The instrumentation associated with twin turbine engines is comparatively 'dense', as this view of the flight deck and main instrument panel on a BN2T variant shows.

Compare that with the classic simplicity of the panel for the 'standard' BN2B Islander, *c.*1981.

In 1991 the company offered a three-blade propeller option, designed to be smoother and quieter in flight. The Hartzell three-blader has equipped a number of aircraft including this Air Wakaya example. Aurigny Trislanders have both type of propeller, three-bladers on the wing engines and a two-blader on the tail.

Even a four-blade propeller was considered. This Hoffman prop was fitted to a BN2T in 1992.

A very striking Islander development was the ducted fan variant produced by Miles Dufour. The aircraft was powered by Dowty Rotol fans inside large ducts and driven by piston engines. G-FANS surprised visitors to the Business and Light Aviation Show at Cranfield in 1977 with its performance and quietness.

Another view showing the remarkable frontal aspect of the fan-powered variant. The Islander proved an excellent test bed for Dowty's quiet alternative to propellers.

In a bid to offer enhanced baggage space, Britten-Norman contrived 28cu.ft of extra room by extending the Islander's nose. This feature was never incorporated in production aircraft because of its effect on the aircraft's centre of gravity, though the longer nose did become standard on the Trislander.

By the early 1990s operators were expressing a need for all the well-known turbine Islander/Defender benefits but with more internal room and greater payload. Britten-Norman responded with the BN2T-4S Defender 4000, a less extensive stretch of the basic airframe than the Trislander and without the third engine. Lengthening the fuselage with a single 30in plug (Trislander has two plugs), using the 53ft-span wing rather than the standard 49ft wing and enlarging the tailplane resulted in an aircraft having a usefully larger cabin than that of its predecessor and almost twice the payload.

Britten-Norman's protoype Defender 4000, G-SURV (denoting surveillance) is seen over the BN factory while on short finals after its maiden flight on 17 August 1994. The type made its public debut on static display at the 1996 Farnborough Air Show. This prototype went on to Police Aviation Services at Staverton in 1998.

The longer forward fuselage and cabin of the Defender 4000, seen clearly in this view, are complemented by a new extended nose able to house a variety of search radars.

A Defender 4000 on flight trials.

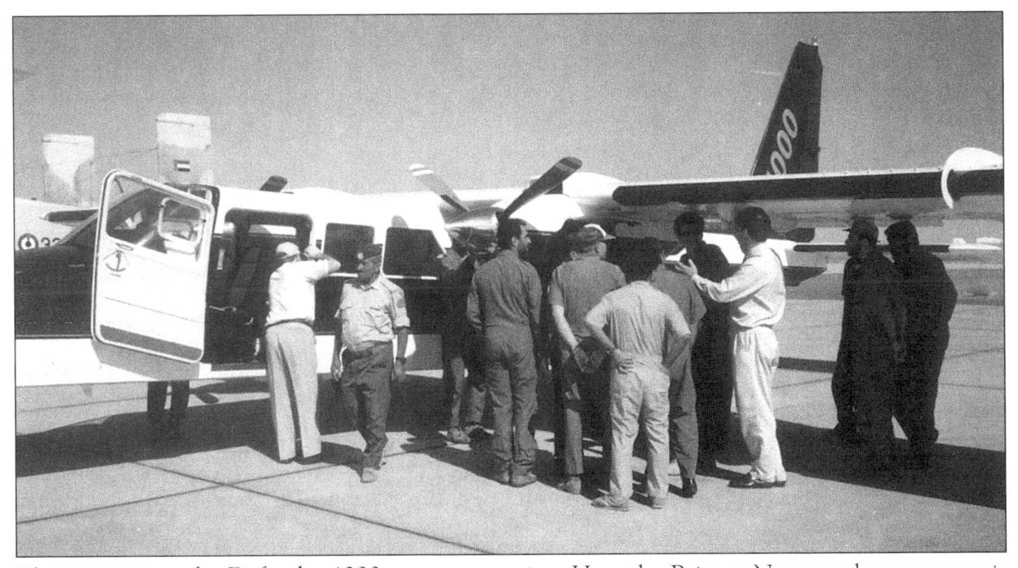

The response to the Defender 4000 was encouraging. Here the Britten-Norman demonstrator is seen attracting attention in the Middle East.

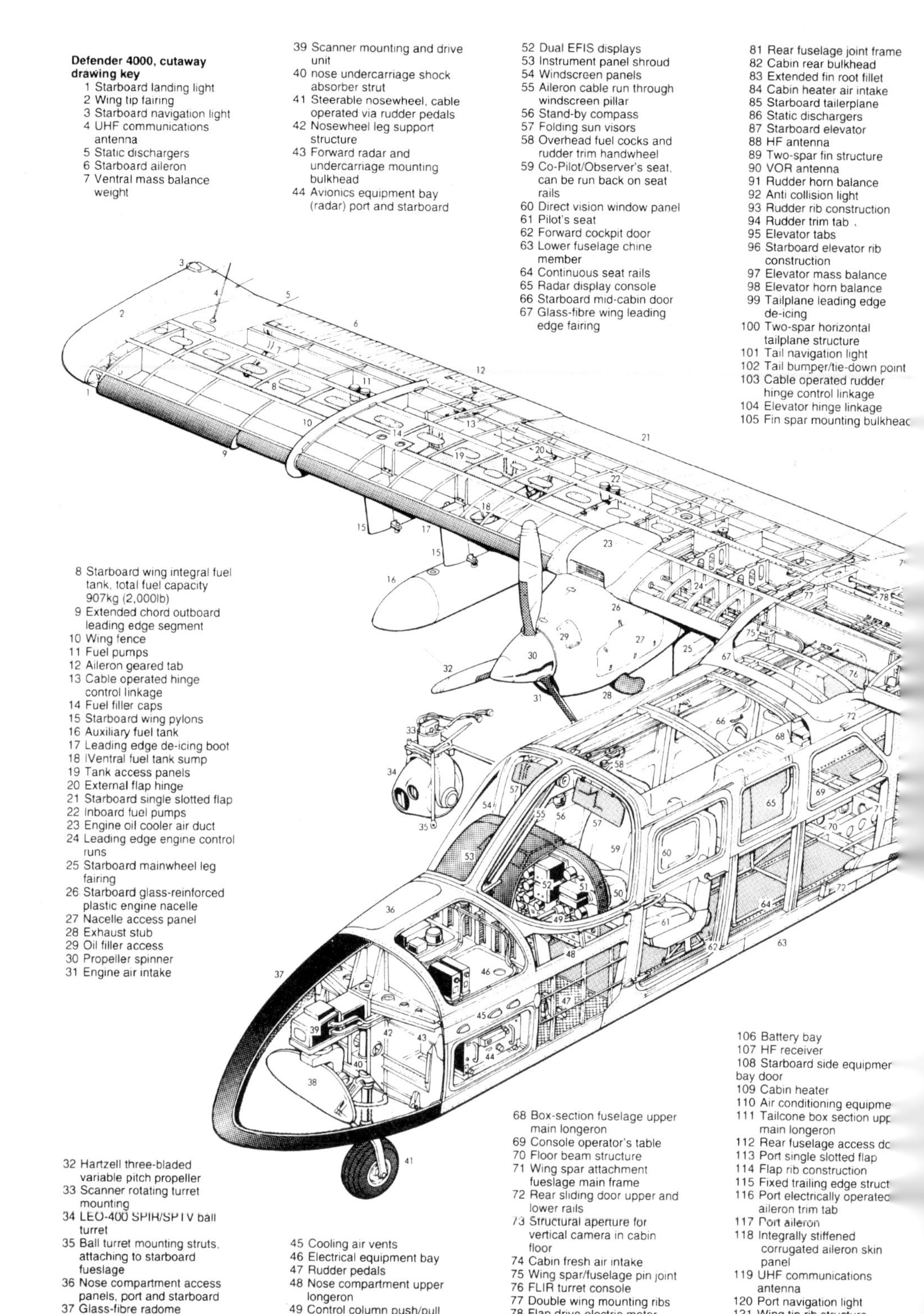

A cutaway drawing of the Defender 4000. Note the Sting Ray torpedo, Sea Skua air-to-surface anti-shipping missile and long-range fuel tank beneath the port wing and a ball turret with

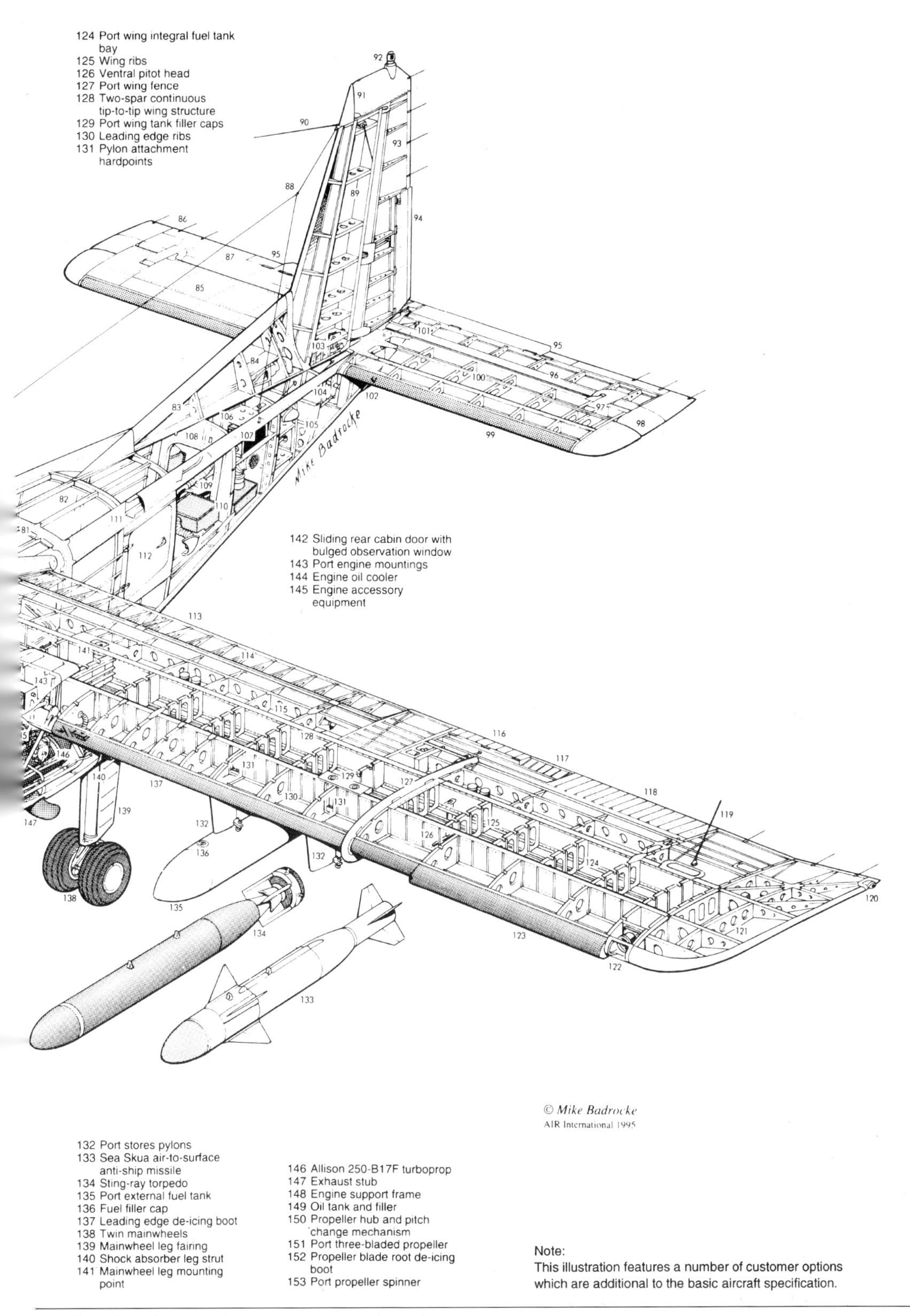

infrared and television sensors to starboard.

Britten and Norman's desire to create a successful light, single-engine, general aviation product, unassuaged by the Finibee, surfaced again during the late 1960s. Boldly deciding to take on established American light aircraft manufacturers such as Piper and Cessna, they produced the BN3 Nymph. The prototype, seen here at Bembridge, first flew on 17 May 1969 and attracted great interest at the Paris Air Show that year.

In a departure from their 'simplicity first' philosophy, Britten and Norman gave their BN3 a streamlined fuselage with big windows and a wrap-round front screen and long, slender high wings which could be folded back to facilitate stowage in hangars. Though the design of the Nymph, seen here near Bembridge, proved sound, plans to produce the handsome four-seater in kit form at £1,930 a copy never went ahead, due to depressed market conditions at the time. G-AXFB subsequently spent years in storage but flew again as the NAC-1 Freelance in 1984, when the project was briefly resurrected at Sandown by a new Desmond Norman venture, the Norman Aircraft Company. The aircraft remained at Sandown where it was, in 1999, in private ownership.

The BN2B, introduced after the Pilatus acquisition, offered higher gross weight, an improved interior design and soundproofing, a fresh air system, a new instrument panel and many customer-related features. It first flew in August 1978 and led to a surge in orders. This example, converted from a BN2A, was used as a BN2B demonstrator and subsequently went to Hawker Pacific, Britten-Norman's agent in Australia.

A later BN demonstrator, registered G-DEMO, was this BN2T Turbine Islander, pictured *c.*1985.

Following pages: Islanders have operated above and below the Arctic and Antarctic circles. The BN2A parked here on a makeshift hardstanding on snow is one of a dozen operated at various times by Munz Northern Airlines, which ordered nine new aircraft in the late 1970s having operated three used aircraft acquired earlier in the decade. Munz flew as far north as 68N, carrying cargo and mail to communities where the Islanders, nicknamed 'the airborne dog sledge', were often the only link with civilization during the northern winter. Munz president Dick Galleher once commented, 'We only stop flying when the temperature drops beyond minus forty-five degrees Fahrenheit.'

Four

The 'Go Anywhere' Islander

With its rugged qualities, forgiving handling and ability to operate away from normal base facilities, the BN2 has been a firm favourite worldwide, including many developing and third-world countries. The type's large, high wing, tough long-stroke undercarriage with twin wheels on the main legs and high power-to-weight ratio enable it to master rough strips of less than 1000ft hacked out of jungle clearings, often with rivers and trees at each end. A Douglas Airways Islander is seen here against the backdrop of tropical rain forest in Papua New Guinea.

The Islander has been a lifeline to many remote villages like this one in Lesotho. The BN2A-21 pictured spent two years serving in the Transvaal before being transferred to Lesotho Airways Corporation in 1979. The sale to the national airline was clinched after an Islander from BN's South African sales distributor had moved 14,000kgs of food into rough landing strips in a mountainous part of the country cut off by floods in early 1974. A pilot for Islander Aircraft Sales later enthused: 'It flew twenty-one mercy missions without missing a beat.'

Remote communities are often pleased to see the Islander and like to gather round. Pilots of this Missionary Aviation Fellowship aircraft in Papua New Guinea tend to park across the uphill end of the crude one-way strip, so reducing the chances of the aircraft rolling down hill should the parking brake be released!

A BN2B equipped for spraying was donated by Live Aid to the Plant Protection Division of the Mali Ministry of Agriculture to help combat locusts.

Three BN2As operated by the Brazilian government's Foundation for Indian Affairs became firm favourites with pilots supporting the Yanomami, a tribe of Indians discovered as late as 1973 in the remote north-east of the country. The pilots described the Islander as 'unique – the only possible aircraft for the job'.

The Islander's unique qualities are appreciated in highly developed countries too. This example, appropriately matched by the 'Jeep' utility vehicle, operated to rough ranch strips in Texas and Mexico with Texan airline La Posada Airways during the late 1960s. According to chief pilot at the time, Byron 'Skip' Reed, 'We regularly use one strip that is 1000ft grass yet we can go in and out at full gross. But the same ship is comfortable on 120mph finals at San Antonio International Airport to expedite traffic.'

Ample reserve power with a light airframe and aerodynamically clean wing enable the Islander to maintain useful load capability in the heat of the desert or at high locations where the air is thin. Here an Islander flies over a dog team on a mountain (from a painting of a photograph).

The British Army's Islander AL Mk 1s proved well up to local conditions during active service in the 1991 Gulf War. The low footprint from the twin wheels on each main landing gear leg prevented the Islander from 'bogging down' in sand. Nor was the aircraft phased by blistering heat, operating successfully in outside air temperatures of 50C. Seven AL Mk1 Islanders are currently operated by the Army Air Corps.

The versatile Islander is equally at home in cold climates. The penguins walking by this example are inhabitants of the Falkland Islands.

The two vehicles sharing this beach in the Falkland Islands epitomize rugged utility. The Islander's tough, go-anywhere virtues have often been compared with those of a Land Rover. This BN2B, manufactured in 1982, is seen in the colours of the Falkland Islands Government Air Service (FIGAS).

A FIGAS Islander being 'guarded' by a bull elephant seal while sharing an East Falklands beach.

The Islander has always appealed to commercial short-haul operators, mainly for commuter, third level and start-up services. In keeping with its name, it has been a lifeline for island communities and has been at home among islands in all the world's oceans, in major archipelagos like Japan, Indonesia and the Philippines and around mainland Europe, North America and Australasia. In the Orkney Isles, off northern Scotland, populations of between 140 and 900 have come to rely on a service started in the late 1960s that links five main islands – Stronsay, Sanday, North Ronaldsay, Westray and Papa Westray. The services, utilizing landing strips little better than farm fields, are rarely cancelled despite the high winds for which the Orkneys are notorious. Here a Loganair Islander approaches Huip airstrip at Stronsay.

Britten-Norman's best seller is also familiar around its home island, the Isle of Wight, off the UK's South Coast. Here the isle of Wight's western extremity is seen under the port wing tip.

BN developed this ski converion to enable New Zealand's Mount Cook Airlines to land tourists on snowfields high up in the mountains. Mount Cook subsequently ordered three ski-equipped aircraft to replace Cessna 185s.

Above and below: Another development in 1974, undertaken as part of a large contract with the Philippines Aerospace Development Corporation, was to make the Islander amphibious by adding floats. An ability to land on water throughout the archipelago as well as using strips on the islands themselves was clearly attractive. Unfortunately, despite extensive design work, the plywood floats produced in prototype form proved too large and heavy, causing the project to be shelved pending development of a more powerful aircraft variant. With hindsight, BN should perhaps have called in float specialists! Full-scale model floats were constructed and mated to the airframe. Note that the aircraft would still have been able to land on its wheels.

By 1994, Islander/Defenders were in service in more than 115 countries and had clocked some 10 million flight hours. This BN2A-26, seen here passing Yarmouth Harbour, Isle of Wight, was later operated in Israel, Greece, the UK and Guadeloupe.

Air Mahe, surely with one of the most beautiful island groups in the world as its 'patch', used this Islander for three years from 1974 before relinquishing it to sister operator Seychelles Airlines.

This BN2A-20 was the aircraft Britten-Norman used for ski development trials in the French Alps, c.1974.

Following page: The Islander has turned out to be a sturdy and tireless workhorse, adaptable to an amazing variety of roles. Parachutists find the BN2 ideal for their purpose. Wide centre of gravity limits and low speed make for stable exits. Wide doors mean that a complete sky diving team, like the Red Devils seen here leaving their own Islander, can exit fast and remain close to each other – important for their dramatic exhibition shows. The high wing and tailplane give safe clearance for jumpers while a high jump rate plus rapid climb and powered descent maximize productivity for parachuting organizations. The Red Devils felt that, when the time came to replace their original piston Islander, the only upgrade/replacement for them was a turbine-engine version

Five

Utility Plus!

Boarding for an exhibition jump.

This interior is equipped to seat four club or trainee parachutists.

Superb handling and downwash from the large high wing facilitate crop spraying, one of the original tasks envisaged by Britten and Norman. This BN2 is taking part in a programme to control pests afflicting date palms in Abu Dhabi. Operating at up to 120 mph, the Islander can cover twenty-five acres per minute in 100ft swathes.

Six specially equipped Islanders helped Desert Locust Control, a joint initiative of Ethiopia, Kenya, Uganda, Tanzania and the Sudan, to fight one of the region's scourges, locusts. The Desloc aircraft had two wing-mounted Micronair rotary atomizer spray units, supplied with chemical spray from an internal tank. The tank, seen in this photo, could carry 1,300lb of chemical. Carrying a full spray payload, the Islander could still take on enough fuel to fly 700 miles. Special modifications enabled the aircraft to fly through locust swarms if necessary.

Above and below: Aerial spraying against pollution has become an accepted technique in many countries. These distinctive Islanders of Harvestair, Southend, UK, in the early 1980s had spray booms fed with dispersant from tanks installed inside the fuselage.

Farms are a natural environment for Islanders, many operators treating BN aircraft much as they do their tractors. Here goods are manhandled, under primitive conditions, into an aircraft of Harold's Air Services, Alaska, apparently in the corner of a field.

Search and rescue is another role for which the Islander, with its high wing, safe low speed and low flying capability, is well suited. This impression of the Channel Islands Air Search Islander, from a painting by Gerald Palmer, shows the aircraft flying over the Roches Douvres.

The Islander can deliver life rafts following an emergency at sea. This aircraft was specifically equipped for the job.

When it was promoted in the air transport market, Britten-Norman's creation filled a gap between light general aviation aircraft and small airliners of thirty seats plus, providing up to nine revenue seats and enabling 'third level' operators to make a profit on low-density, short-haul routes for the first time. BN's distributor for the Americas, Frank S. Jonas, said, 'The Islander has the lowest capital cost per revenue seat of any aircraft intended for operation on low-density routes. It is ideal for maintaining links with the numerous small towns and villages found in all countries.' This example is from Munz Northern Airlines.

With its 10ft long, low, unobstructed floor, wide doors both sides and sizeable 'box' interior, the Islander has the versatility to carry passengers, cargo or both. The adaptable workhorse can deliver a ton of urgent cargo, safely secured to built-in tie points, in the morning and quickly revert to seating to carry nine passengers in the afternoon. Being loaded here is an Islander belonging to Falcks Redingskorps Odense of Denmark.

Marketers claimed that an aisle would be redundant on the Islander as the low cabin floor enables passengers to board easily from ground level, without steps. The generous centre of gravity range means that disposition of load raises few difficulties. Interior fits range from standard 'utility' bench-type seating, as in this example...

...to luxurious leather seating along with superior trim and curtains, as seen in this executive interior.

This BN2 passenger interior, implemented by Isles of Scilly Skybus, provides a comfortable cabin environment for a full Islander complement of ten. The seats are orange while the cabin trim is light beige.

The Islander is widely used as a personal aircraft or executive/VIP transport. Executive interiors can feature air conditioning, curtains, luxuriously upholstered seats, high quality acoustic suppressing trim and – as in this early example - a drinks dispensary.

The Islander has helped open up the interior in countries where alternative transport is hazardous and unreliable. This Air Gabon BN2A was one of three providing a link between the capital Libreville and communities deep in the heart of Gabon during the annual rains, when local roads become almost impassable. It also ran a scheduled service between Port Gentil and Gamba for Shell Petroleum. Air Gabon also had a Trislander.

The relatively flat fuselage sides of Britten-Norman's aircraft make them suitable as 'flying billboards' as well as effective passenger and freight carriers. This Trislander, actually belonging to Channel Islands operator Aurigny, raised the profile of Hambros Bank.

The ability of Britten-Norman's legendary fixed wing product to carry out 85% of the missions that a helicopter can, at about a third the cost, has always appealed to police and other law enforcement agencies. High wing, safe slow flying and twin-engine assurance enable the type to loiter over urban as well as rural areas while observers note and record activities of interest. This BN2T turbine Islander, operated by the Dutch National Police, constitutes an 'eye in the sky' in the war against drugs, for example, detecting and monitoring the movements of suspicious small marine vessels and light aircraft.

Seen here is a typical simple console/workstation for an observer in the back of a police Islander.

Islanders and Defenders are in service with the police forces of the Netherlands, Belgium, Ireland, Cyprus, Oman and Malawi as well as the UK. This BN2B Islander is seen over Hampshire, whose police force was a pioneer of fixed-wing air support in the UK. An order for a turbine-powered Defender 4000 was announced in 1999 to replace this aircraft after years of successful service. The specified equipment fit includes multi-band communications, weather radar, TV, thermal and photographic sensors and full navigation aids. A loud hailer and searchlight can also be fitted.

Police, coastguard and similar operators commonly have infra red sensors. Here two ways of mounting a sensor turret are evident: on the fuselage, as with this BN2T coastguard Defender…

…and on an under-wing hardpoint, seen here on a BN2T Maritime Defender.

The Irish national police, or Garda, were the first police force to operate the Defender 4000, the roomier cabin of which permits the installation of sophisticated observer consoles like this.

The Falkland Islands Government Air Service used the BN2B-26 in a maritime law enforcement role, particularly fishery protection. Here a patrolling aircraft investigates an unflagged, unidentified vessel. Another duty was plotting the passage of nuclear waste ships through the islands' waters.

Morocco is the world's largest operator of turbine Islanders, with fourteen BN2Ts serving with the Ministry of Fisheries. The aircraft patrol a million square kilometres of the economic exclusion zone (EEZ), out to the 200 nautical mile limit, in order to protect the country's rich fishing grounds. Pleased with the performance and low running costs of the BN2Ts it already had, Morocco ordered four more in 1993.

Photographers appreciate the superb wide unobstructed view secured by removing Islander's rear door. The high wing is out of the way visually and slow flying means that repeated shots are available before a scene is overflown.

In this variant, photographers or film makers can operate from a seat that becomes a working couch when fully reclined, helping the photographer to hold the camera steady while shooting through the open side door.

Fixed camera installations have been extremely varied. This bulky unit is for high-resolution photographic survey.

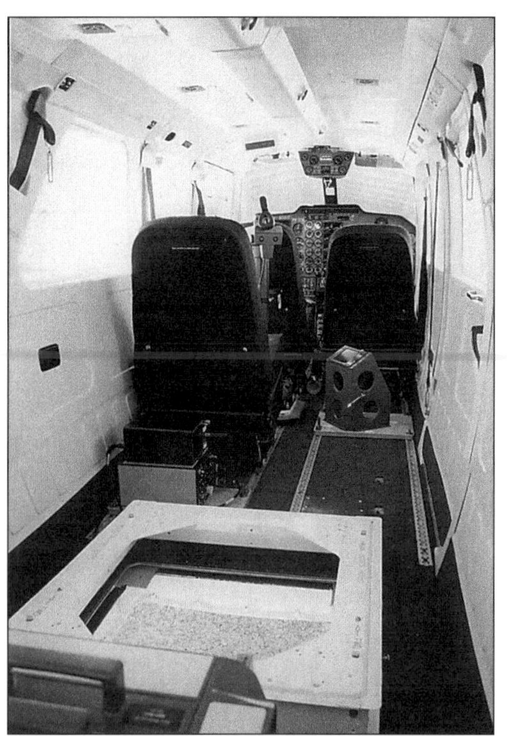

Britten-Norman have also developed an optional 'camera floor' modification for police and other law enforcement users. This example, however, is in the floor of a commercial Defender 4000 used by Sabah Air for aerial survey photography.

The Islander air ambulance has room for three stretchers and an attendant, or two stretchers plus attendant and one or more 'walking' cases or relatives. Its STOL characteristics and ability to land on anything from a football pitch or golf course to a road means that it can get close to an accident or disaster, saving time and lives. Here a patient is embarked into an Islander of Loganair, well known for its air ambulance lifeline to the Scottish Highlands and Islands.

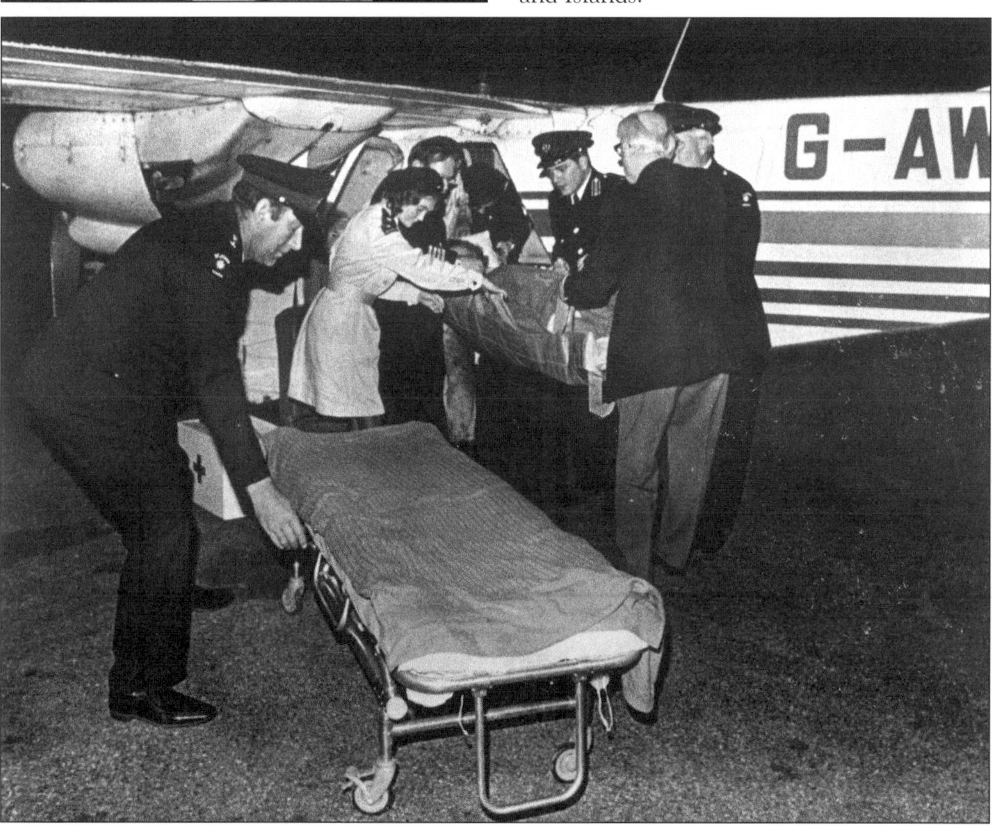

This two-stretcher plus attendant layout illustrates the cabin's practical utility in the air ambulance role.

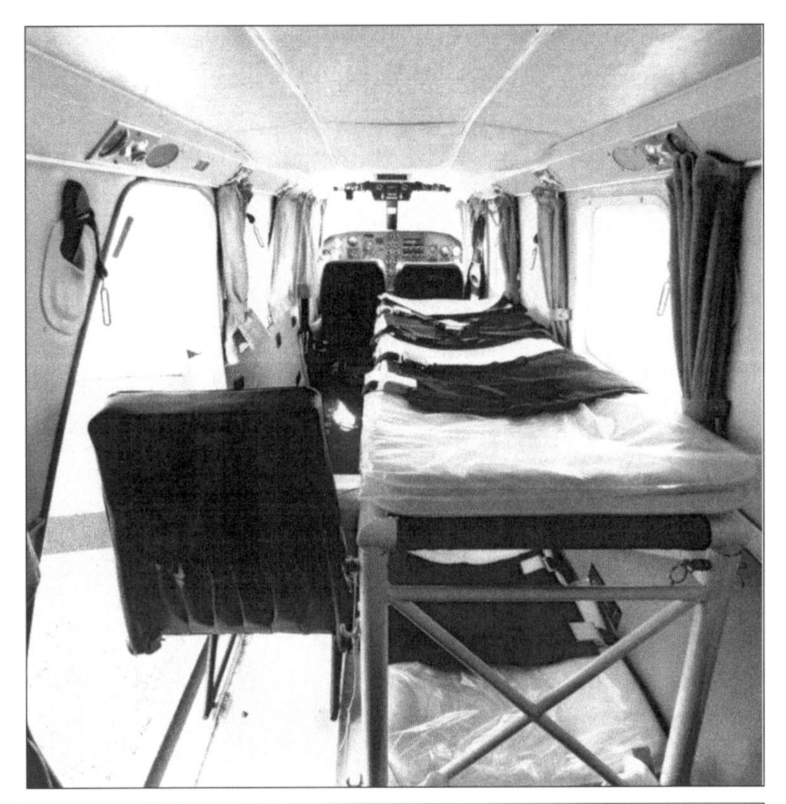

An aircraft which can take off in 520ft at maximum weight, clear a 50ft barrier in 1,020ft from brakes off, and land slow and short in 400ft is a boon for maintaining unscheduled and emergency services to remote communities. This has certainly been the case in East Africa, for example, where five Islanders operating with the Zambian Flying Doctor Service in the early 1970s were so much admired that UK Crown Agents promptly ordered another air ambulance variant for Zaire. The East Africa Flying Doctor Service also operated the type from Nairobi from about 1975. In a similar role much further south, this aircraft is a BN2A of the Servicious Medico Aereo of Lourenco Marques, Mozambique, pictured during the mid-1970s.

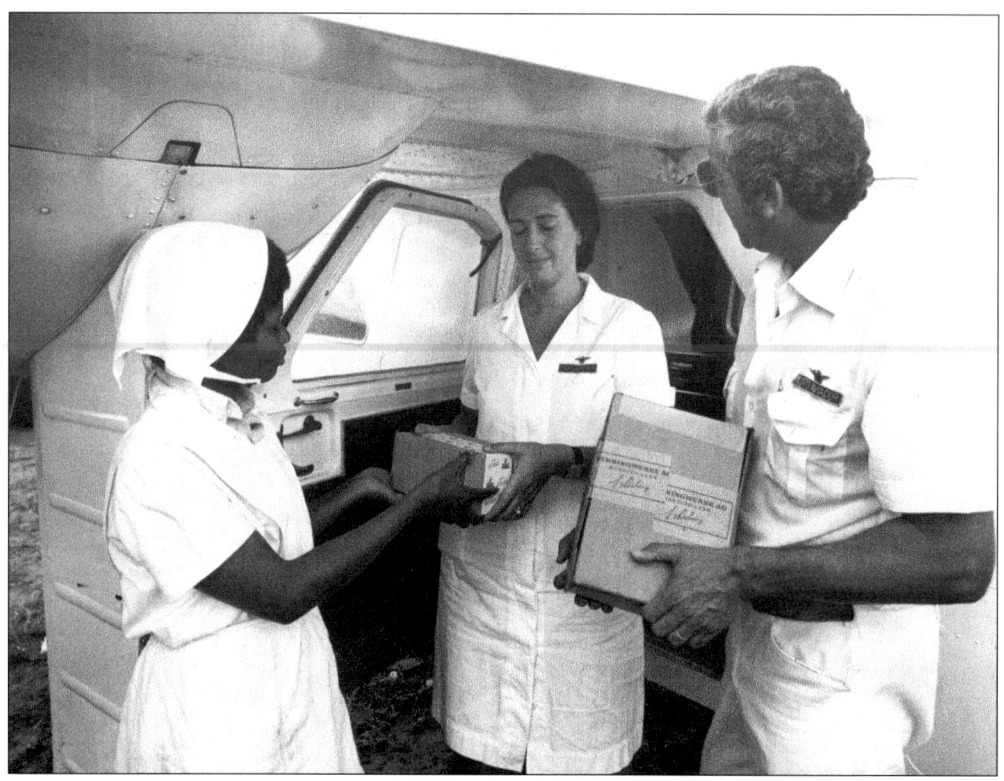

Handing out medicines at an outpost in Dafuri, Mozambique, 1974.

Sensing exploitable resources from the air has accounted for several protuberances from the amenable Islander airframe. Scientific geophysical survey instruments such as magnetometers are carried in probes, as on *Romeo Papa*, used by ENCAL of Brazil to hunt for minerals in the country's extensive hinterland. The tail-mounted 'sting' can detect magnetic changes which indicate the presence of minerals in the earth's crust.

In a water bomber variant developed by PBN's distributor in Argentina for fighting forest fires, a carefully engineered tank in the cabin could dump its 810 litres of retardant onto a fire in just two seconds. The conversion kit for the FAA-certificated Islander could be removed within hours, making the aircraft available for normal duties again.

Few Islanders are used as church and altar but this example took matrimony to new heights when John and Michelle Buckland were married 5000ft up over Itimpi near Kitwe, Zambia, in 1990. Pictured still on a 'high' are the bride and groom with the Reverend Jim Hess, who married them, pilot Major Julius Mbewe, who flew them, and young relatives.

Strangest of all was this system for sensing radioactive trace elements as an indicator of soil type for wine growers. The arrangement of linked nose and tail probes with wing-tip sensors made this aircraft instantly recognizable.

Six
Islander as Warrior

With Islanders succeeding so well in the law enforcement role, it was inevitable that the rugged dependability and useful payload of the platform should appeal to the military also. This was even more the case after 1971 when Britten-Norman developed a 'hardened' Islander which they called the Defender. A strengthened airframe and four underwing hard points enabled the new variant to carry an impressive range of military hardware. This Army Air Corps BN2T Defender (Ministry of Defence designation Islander AL Mk1) flew British Forces Middle East (BFME) officers between base and forward units during operations in the Gulf in 1991. Forces personnel thought the aircraft had plenty of personality and called it 'Pinky'.

The sight of the ordnance and other devices exhibited with this Defender on air show duty would be enough to convince anyone that BN's new type could be a tough and gritty warrior. Bombs, rockets, gun pods, sensors and long-range fuel tanks could all be carried under the strengthened wing.

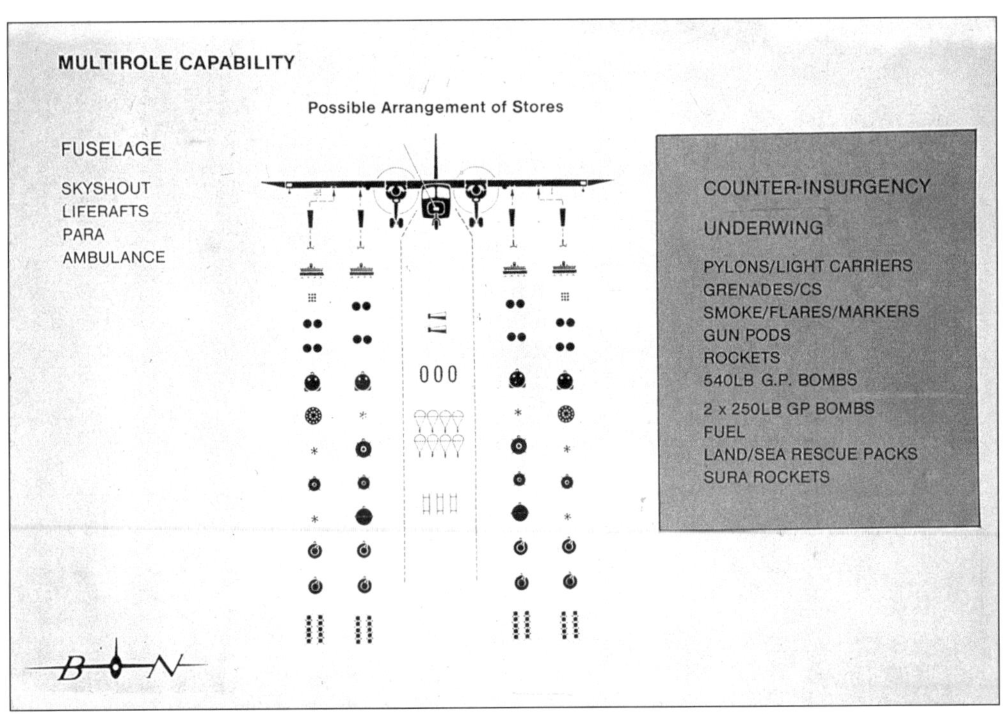

This drawing from a BN brochure emphasizes the stores flexibility of a counter-insurgency Defender.

Above and below: The Defender can turn aggressive when it has to. Machine guns can be pod mounted under the wings or installed in the cabin as these examples show.

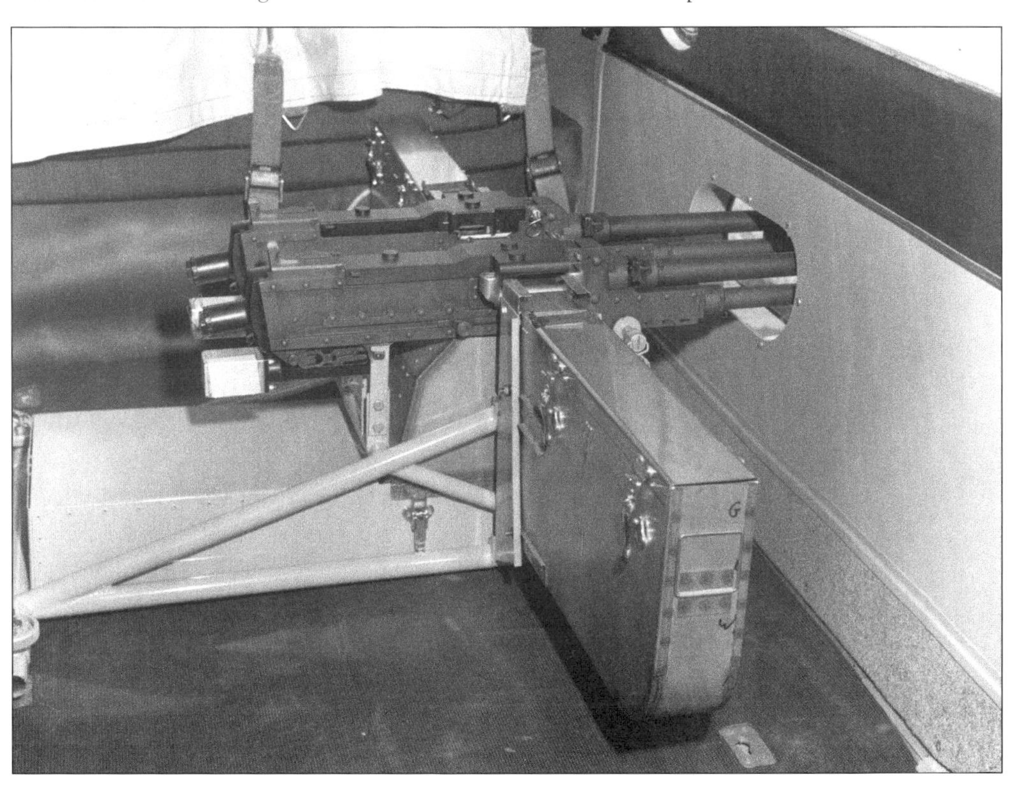

This BN2B is firing SNORA rockets during trials.

Angry! The Islander/Defender can be used in a counter-insurgency role, as the underwing rocket cluster on this Mauritania Islamic Air Force example suggests.

This flare countermeasures pod constitutes an effective defence against missiles.

Lifting trials with an AL Mk1 (British Army designation) clearly showed that downed Islander/Defenders can be recovered by twin-rotor Chinook helicopters when required.

The turbine Defender offers military users the performance and payload benefits that have so appealed to commercial users. The first Maritime BN2T was delivered in 1982 and since then the variant has sold to over twenty countries. The Maritime Defender can typically carry a 360 degree search radar, forward looking infra red, magnetic anomaly detector, sonobuoys and Omega/GPS navigation. To pack a punch against surface vessels or submarines, the aircraft was also offered with four Sea Skua anti-shipping missiles, depth charges or two Sting Ray light torpedoes. An electronic countermeasures or electronic support measures pod can be installed. This example was purchased by the UK Ministry of Defence in 1986 for Sting Ray carriage and release trials.

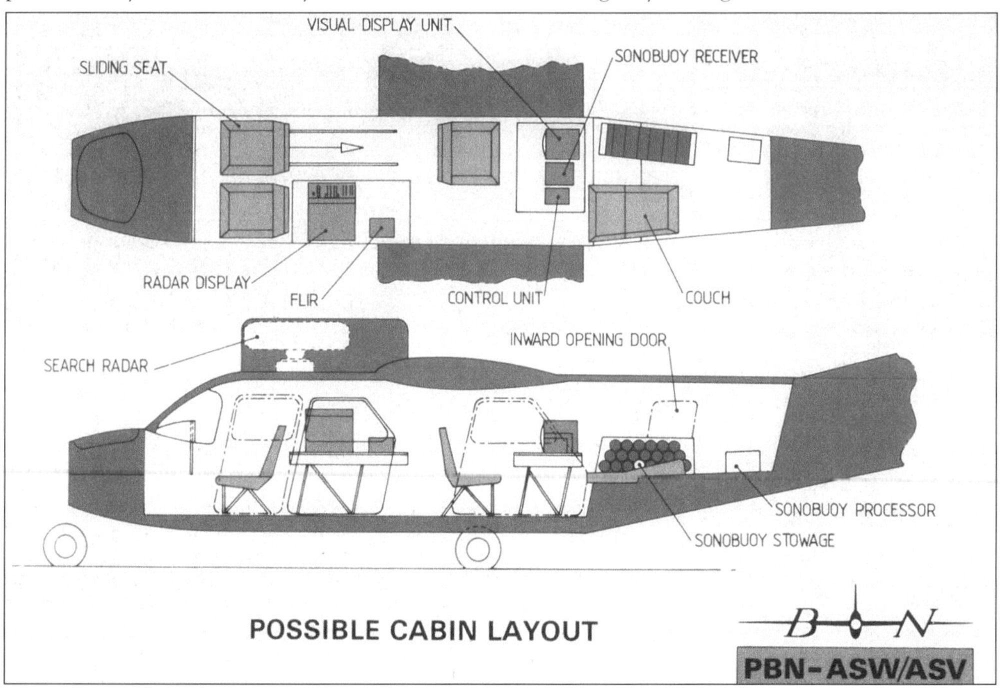

The interior layout of an ASV/ASW Maritime Defender.

With Pilatus' encouragement, Britten-Norman worked hard during the 1980s to develop further military variants, with the emphasis on remote surveillance and intelligence gathering. A series of distinctive bulbous and 'platypus'-nosed aircraft were clearly advanced, military and 'hush-hush'. In each case the nose housed a high-performance radar which was at the cutting edge of contemporary technology. The first to fly, in 1984, was this Corps Airborne Stand-Off Radar (CASTOR – later ASTOR) variant developed for the Army and Royal Air Force. The Ministry of Defence selected BN2T as the trial platform because it was affordable, easy to maintain, could fly in all weathers and could use makeshift airstrips. It was fitted with an advanced Ferranti radar designed to monitor battlefield movements from a distance. Note the unique 'platypus' or 'metal detector' nose.

Chief test pilot John Ayers and observer Peter Ward are seen with the CASTOR aircraft after the May 1984 first flight.

Many of the CASTOR project team are seen here with the 'Ferranti aircraft'. They are, from left to right, production director David Fear, technical director Bob Wilson, production engineer Richard Bishop, flight test engineer Ian Wightmore, chief inspector Jack Griffen, stress engineer Terry Neary, chief draughtsman Mike Dore, production engineers Keith Winter, Peter Williams and Peter Sothcott, flight test engineer Peter Ward, test pilots John Ayers and John Davies and Jim Roberts from avionics.

Subsequently, a second modified turbine Defender flew with another radar, the Thorn-EMI Skymaster, in a joint project to provide airborne early warning against low-level attack by missiles or aircraft. Of all the various attempts at AEW, the Pilatus Britten-Norman/Thorn EMI Defence venture was said to be the easiest on defence budgets, earning a re-rendering of the AEW acronym as 'affordable early warning'. The AEW Defender first flew in July 1984 and was launched as a fully operational system in March 1987.

The nose radome housed, as this view shows, the 360 degree Skymaster radar which, with its advanced pulse Doppler processing, could track up to 250 targets, fast and slow and at high or low level. Note the ease of access afforded once the lower radome cowl was detached.

Observers in the AEW Defender's cabin obtained situational awareness at consoles like this.

A large nose was again useful for the BN2T-4R Multi-Sensor Surveillance Aircraft (MSSA), resulting from an agreement in 1990 between Pilatus Britten-Norman and Westinghouse Electronic Systems of Baltimore, Maryland. Inside was a powerful Westinghouse multi-mode radar developed originally for the F-16 fighter. The cabin was crammed full of sensors and electronics, leaving conditions for observers rather cramped. This was one of the triggers for the subsequent development of the Defender 4000 with its enlarged fuselage and cabin.

A UK Type Certificate was awarded in November 1995 and significant orders were predicted. However, the MSSA was overtaken by events and eventual decisions by several countries to acquire larger, albeit more expensive, surveillance platforms put a shadow over this promising concept. One of four MSSAs acquired by Westinghouse is seen here over 'home territory' at Baltimore, USA. Note the fillet added forward of the fin to enhance longitudinal stability and the double wing fences.

Another venture by Pilatus Britten-Norman into the high value added military field was this electronic intelligence (Elint) Defender, which had Racal Kestrel electronic support measures installed in underwing pods. In its marketing, Pilatus Britten-Norman constantly stressed that the compact nature of modern electronics means that an airframe of the Defender's size can now carry highly capable mission suites, whereas previous electronic generations would have required much larger aircraft. The Elint Defender was launched officially in 1988. The underwing pod houses the Racal equipment designed to identify and pinpoint hostile radio and radar emitters, whether airborne or on the ground. This aircraft, previously with the Rhine Army Parachute Association, was used to demonstrate the concept.

During the 1990s the US Army used a BN2T – converted from a piston-engine aircraft – as a flying laboratory for night vision equipment. Here Captain Kevin S. Noonan, commander of the US Army Night Vision Laboratory's aviation detachment, accepts the re-engined Islander from Major General Bill Gorton, US Air Force (retired), President of Pilatus Britten-Norman USA.

Seven

Operators

Back at the start, the first production Islander went to GlosAir of Staverton, Gloucestershire, in 1967. GlosAir ordered the aircraft without benefit of a trial flight, so sure were they that the Islander was 'right for the job'. They went on to form Aurigny Air Services to connect Alderney (Aurigny) with Jersey and Guernsey initially and later with Cherbourg and Dinard in France and Southampton in England. The first of twelve Islanders the airline went on to operate was G-AVCN, which inaugurated scheduled services in March 1968. In October 1971 Aurigny also became the world's first scheduled Trislander operator and has since operated up to ten of the unique three-engine aircraft. Note the stylized Alderney lion on the tail of this Aurigny Trislander.

Previous page: This 'cheeky pup' seen alongside Concorde at London Airport is an Islander with one of the world's 'blue chip' airlines, British Airways.

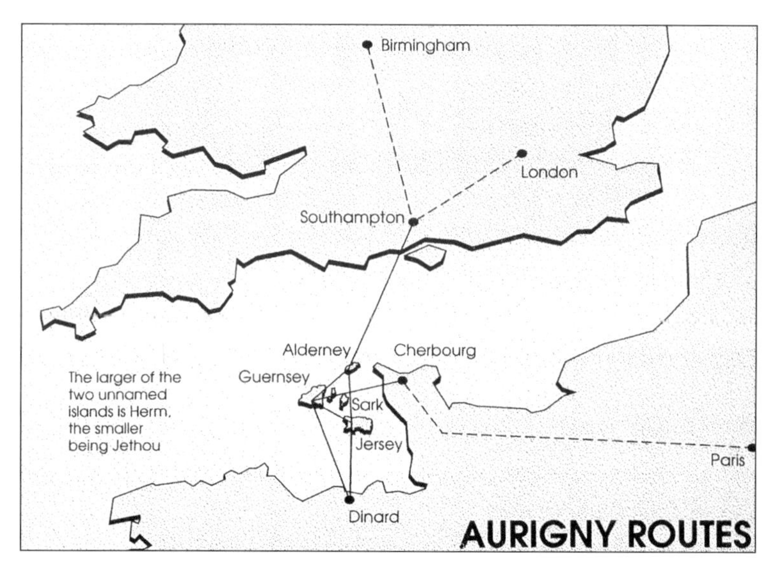

These were the routes Aurigny was flying by the1970s. With a preponderance of short-sector, inter-island work, where engine cycles build quickly, the airline has always preferred piston engines since these are economical to maintain and overhaul.

The other joint launch customer was Scottish commuter airline Loganair. Linking the Scottish mainland with the islands off its north and west coasts, Loganair tested the Islander's short-sector capability to the utmost. Its Orkney services, started in 1967, included the world's shortest flight sector, a two-minute hop from Westray to Papa Westray. Even the longest, between the capital, Kirkwall, and North Ronaldsay is only a fourteen- minute flight. Loganair Islanders were landing up to twenty times a day, often on primitive mud or grass airstrips where natural hazards included grazing animals and cowpats! Here a Loganair Islander makes a low-level sightseeing pass of the 'Old Man of Hoy'.

A philatelic cover commemorating the first scheduled flight to Fair Isle, where Loganair pilots were challenged by a precarious narrow strip 500ft up on a hillside. (Reproduced from *Island Pilot* by Capt. Alan Whitfield, published by The Shetland Times)

The nine-passenger Islander became the smallest airliner in the British Airways fleet in 1994 when Loganair became a BA franchise partner, causing the Islanders to be re-liveried in British Airways Express colours.

A rather different island environment, the tropical islands of the East Caribbean have always been a natural home for Islander/Trislander aircraft, which have played an invaluable role in linking the scattered communities of the region. Many West Indian operations have a simple, unpretentious style. They require an affordable 'start up and go' workhorse, able to withstand the tropical sun, fly many times a day on short sectors and offer twin-engine assurance for sea crossings. Piston engines are favoured for easy maintenance and Avgas burning. Here, while two passengers shelter under the wing of an Air Guadeloupe Islander, another braves the tropical rain. This French Antilles operator acquired two of the type in the mid-1970s and later increased its complement to four.

Antigua-based Leeward Islands Air Transport (LIAT) was founded by Frank de Lisle, who also introduced the BN2 to the Caribbean region, where it remains a mainstay. LIAT purchased six Islanders, primarily through Jonas Aircraft, between 1970 and 1980. Pictured is one of LIAT's examples seen with a Twin Otter and a Dash 8 at VC Bird Airport, Antigua.

Seen undergoing maintenance in Antigua, British West Indies, in 1995 is this BN2 operated by the Dream Company, otherwise known as 'K Club'. Dream has flown hundreds of guests to the resort complex on the idyllic island of Barbuda, known for its dreamy topical beaches and lagoons.

Another specialist in flying guests to smart resorts, Frenchman Claude Deravin, purchased this used Trislander for his St Barts Plus operation in the mid-1990s. The aircraft chosen had to be able to manage St Jean airstrip on the French island of St Bartholemey, surely the most terrifying in the Antilles. Pilots have to drop in to the ultra-STOL strip over a high hill practically on the threshold, then stop quickly before they get to the sea at the other end!

The author flew to Montserrat on this Islander before the island's volcano triggered an exodus by threatening to blow its top and covering half the island with ash. Montserrat Airways' scheduled shuttle service during the mid-1990s used two ex-Aurigny BN2A-26s to cross the few miles from Antigua thirteen times a day.

Another classic Islander operation is the air bridge between the two islands of the Federation of St Kitts and Nevis. Nevis Express, founded in 1993, was soon shuttling across the straits, a six-minute journey each way, up to fourteen times a day using two Islanders. A third Islander augmented the fleet in 1997.

Colour schemes for airlines in the tropics are largely white to reflect the sun – especially when aircraft are 'hot soaking' out on airport tarmacs. Inter Island Air Services, which took over four ex-LIAT Islanders around 1987, tempered its white with an eye-catching livery in bright colours.

Trans Jamaican Airlines of Montego Bay had some five Britten-Norman aircraft during the 1970s and 1980s, of which this Trislander was one. To lend its aircraft some exotic European appeal for the local clientele, the airline gave its aircraft names such as *Miss Manchester* and *St James*!

Oceanair Line of Vieques, Puerto Rico, leased two BN2A Islanders in the late 1970s and early 1980s from Air Investments of Florida. The service connected San Juan and Ponce on Puerto Rico with St Thomas in the US Virgin Islands.

Britten-Norman's 'island hopper' has long been a mainstay among the scattered island groups of the vast Pacific Ocean. Here, in true South Pacific style, Fiji's Air Wakaya 'blesses' its new BN2B in a traditional ceremony involving song, dance and this all-round drape. Air Wakaya had the first of the three-blade propeller variant. Its aircraft also had club-style leather seating, air conditioning and bubble windows to assist tourist viewing.

Another South Pacific operator, Aviazur of New Caledonia, had this unique livery reproduced on the BN2T it purchased in 1996. The *serpent tricot raye* (striped jersey snake) is a good luck mascot. The turbine Islander complemented a piston-engine BN2A supporting the islands' leisure and tourism sector as a commuter aircraft and air taxi. Aviazur wanted an aircraft able to carry a full nine-passenger payload in high temperatures, use short runways and fly routes up to 250 miles. The replacement aircraft also has to serve, on occasion, as an air ambulance.

Air Moorea, an Islander operator since 1968, has had eight of the type. These two BN2B-26s, built in 1989 and 1990, fly from the 'Pearl of the Pacific', Tahiti.

Away from islands and the sea, the Islander has also found an invaluable role reaching the interior of countries where other forms of communication may be scant or unreliable. Here, equipment flown in by Lesotho Airways is set out ready for alfresco field surgery. Lesotho, with three BN2As during the 1970s, has frequently used its Islanders in a humanitarian role.

The United Nations found its Islander, purchased in 1970, more flexible than some of its member states! UN Development Programme managers appreciated the ability to change quickly between passenger, cargo and mixed use. This aircraft helped support UN projects in Sudan.

The UK's Ministry of Agriculture, Fisheries and Food has relied on Britten-Norman reliability in combating illegal fishing in British waters. A turbine Islander operated for MAFF by FR Aviation and affectionately known as 'Noddy' at one time clocked around 1000 hours a year patrolling waters around England, Wales and Ireland – most of it below 500ft! Despite constant low-level operation in a salty maritime environment, few corrosion problems were reported on the BN2T. Note the range-extending underwing fuel tanks and the radar nose.

This ex-Aurigny aircraft, seen here over Sydney Harbour, was sold via Britten-Norman to the company's Australian distributor, Islander Aircraft Sales, in 1973. Within a year it was flying with Pacific Resorts of New South Wales.

The longevity of the Islander has ensured multiple operators for many of the over 1,200 built. This BN2A, which first flew in 1969, journeyed during its career to North America, via Jonas Aircraft, New York, to the Caribbean, where it served with LIAT and Four Island Air Services, then back to the UK, where it joined Air Atlantique in 1987 and shortly afterwards, Air Alba based at Inverness, Scotland. It later returned to Air Atlantique and then became privately owned in Scotland.

Aerial Tours of Papua New Guinea was one of the earliest and most enthusiastic Islander operators in the Far East, having taken delivery of its first in 1968. *En route* to delivery, this BN2A finished first overall in the London-Sidney Air Race of 1969, flown by Capt. W. Bright and F. Buxton – a clear demonstration that sheer reliability gets results!

Malaysia's Sabah Air was the first commercial customer for the BN2T-4S 'stretched' Islander, otherwise known as the Defender 4000. Sabah wanted an aircraft with significant range and endurance as well as the STOL ability to use the many short airstrips dotting the countryside of Sabah and East Malaysia. A camera floor modification was fitted and the aircraft is used for aerial survey and photography.

The Islander can operate reliably in hot climates. Precautions against the direct sun in the daunting climate of the Gulf are clear in this shot of an Emirates Air Services aircraft.

Indonesia, with its 13,000 islands, is an archipelago well suited to Britten-Norman's short-haul excellence. At one time twenty-two Islanders and Trislanders served with ten operators, including Dirgantara Air Services with a fleet of eight BN2s in service in 1997, Djakarta-based Indonesian Air Transport, Bouraq, Indonesian Aviation, Bali Air and PT Inti Indorayon Utama. Seen in the colours of the last named in 1993, this aircraft is at Propat airstrip prior to carrying out an aerial survey of reforestation projects in the district.

Air Liberia acquired three BN2A Islanders and two BN2A MkIII Trislanders in the mid-1970s.

Air Seychelles was unusual among operators in acquiring Trislanders first, in 1976 (the enterprise was then called Inter Island Airways), and Islanders second, from 1977. The operator built up to a fleet of two Trislanders and three Islanders – one of the latter being seen here.

One of Air Seychelles' two Trislanders prior to its delivery in 1976.

The eighteen-place Trislander development of the basic Islander airframe also appealed to Sunbird Air Charters who operated a BN2A during the early 1970s then took a step up in capacity with this Trislander in 1972.

Both Loganair and Aurigny, who pioneered BN2A Islander operations back in the 1960s, became keen Trislander operators too. This Aurigny 'mini airliner' contrasts with the massive 'jumbos' it is keeping company with.

Islander utility has proved its worth not only in the less developed countries of the third world, but in the sophisticated first world too. Indeed, the strongest markets have been in places like the United Kingdom, the United States of America, Europe, Scandinavia and Japan, where second and third-tier operators have taken enthusiastically to the Britten-Norman formula. To the group of Norwegian taxi drivers who formed Teddy Air in 1990, this Islander was another 'taxi', carrying passenger around Oslo in the early 1990s. They leased it from a private owner, who eventually sold the aircraft on to interests in Finland.

The white top on this Euroair Transport Islander contrasts strongly with the dark sea below. This operator was based at Biggin Hill in Kent, UK.

Luftverkehr Friesland Harle, who have been flying tourists to the Friesian Islands off Germany's north coast since 1983, have found the Islander well suited to the rigorous conditions there. Herr Jan Brunzema, director – seen here accepting a BN2B-26 from BN sales manager Rod Lanning in 1990 – commented at the handover, 'Inter-island flying in the Friesians can be challenging. Often the cloud base and visibility are low. Chock-to-chock flight times can be less than five minutes and landing strips are short. Islander, with its excellent STOL characteristics, is about the only aircraft that can land in Helgoland, where the runway is little over 250 metres long.'

A neighbouring airline also serving the Friesians, Frisia Luftverkehr Norddeich (FLN), experienced high demand for its BN2B services, at one point moving 2,722 people in just three days between Juist Island and the mainland. On one notable day, 21 April 1990, six Islanders – including examples loaned by neighbouring operators – carried 1,542 passengers.

Britten-Norman's rugged short-haul transport has much to offer the group of four major and some 3,000 off-lying islands, many of them well populated, that constitute Japan. Residents of islands served from Nagasaki have come to depend on the familiar Islanders of Nagasaki Airways for their standard of life. As well as carrying passengers, the aircraft provide an emergency medical service, transport blood and fulfil other vital services. The local airline operates five aircraft over sectors of forty to eighty nautical miles and at frequencies of up to twenty-eight round trips a day.

This striking red and black-liveried BN2B-20 entered service with Kyokushin Air in 1995. The inspiration for the livery was the crested ibis, a bird of special cultural significance in Japan. The Islander flies commuters between the north-west coastal town of Niigata and the off-lying island of Sado.

Japan's mighty Honda organization used their Islander for transporting demonstration personnel, sometimes with their motorcycles, around Europe.

European enterprises, too, have appreciated the Islander as a productive corporate utitlity. Volkswagen used this example, delivered in 1972, to run an express parts service between the German factories and customers in other countries. One commentator described the aircraft as 'the Volkswagen of the air'.

Numerous law enforcement agencies and armed forces have favoured the Islander and its 'hardened' derivative, the Defender. Some 95% of all turbine Islanders sold by Britten-Norman are to police, coastguard and other law enforcement operators. In the UK, Police Aviation Services (PAS) provides air support for any UK police force that requests it, though some forces have their own. Here BN aircraft operated by PAS, the Hampshire Police and the Cheshire Police are seen at the PAS base at Staverton, Gloucestershire.

The Royal Ulster Constabulary in Northern Ireland achieved effective air support with its BN2T turbine Islander during the 'troubles'.

Few would wish to argue with the Cyrpus Police when faced with this BN2T armed with underwing machine-gun pods.

This photograph of an Omani BN2 flying low alongside a Fairey-built fast vessel during the 1970s seemed to symbolize the combined air and marine background of the Fairey Group. Number 5 Squadron of the Sultan of Oman's Air Force has had good use from eight BN2A-21 Defenders delivered in 1974. The aircraft operated mainly as troop and supply transports, though some had provision for stretchers and three had VIP interiors. Commercial operator Gulf Aviation took two Islanders in the early 1980 and the Omani Police had one BN2T.

Another enthusiastic user of Britten-Norman products is the Belize Defence Force, which took delivery of two BN2B-21s in 1982. One task for the Defenders has been to make daily sweeps of isolated areas of jungle looking out for small clearings where drugs are grown.

A prestigious mission for this Belizean Defender was to carry Queen Elizabeth II as a passenger during a visit by Her Majesty to the ex-British dependency.

Many operators are based in archipelagos – land masses and countries made up of dozens, hundreds or even thousands of islands – where small aircraft are essential for maintaining effective communications. The Philippines have been an intensive user (and builder) of Britten-Norman's classic. Islander/Defenders were put to good use by the Philippines Air Force (the country's largest user, having taken some twenty-two of the type), Navy and Constabulary as well as the National Development Bank and National Oil Company, both of Manila. Built by the Philippines Aerospace Development Corporation (PADC), over a dozen went to Philippine Aerotransport and nine were delivered to Philippine Airlines.

Surveillance Australia operates six BN2B-20 Islanders with the Australian Customs 'Coastwatch' programme. A special observation window, just visible behind the undercarriage strut, was fitted to improve downward visibility for observers in the rear of the aircraft, especially during surveillance at low altitude.

This view shows an Oman Police Defender quite at home under desert conditions.

Dubai is another Gulf State to operate the BN2T Defender, as this UAE Air Force example shows.

The Indian Navy, after two successful decades with a fleet of six Defenders which first entered service in 1977, decided to give these aircraft a new lease of life. With design and engineering support from Britten-Norman, it therefore stripped the BN2As, fitted them with 400hp Allison 250-B17C gas turbine engines, and thoroughly renovated the aircraft. BN developed the modification scheme and sent engineering teams to support the Indian Navy as the work progressed at the Cochin naval base.

The Rhine Army Parachute Association, a faithful user of Islanders for about two decades, operates them intensively from its airfield at Bad Lippspringer near Dortmund. RAPA has made through-life costing into a fine art, managing to replace each aircraft just before its maintenance costs start to spiral. In 1997 it ordered its fifth turbine Islander in a row. These followed two previous piston-engine versions, the first of which was delivered in 1981. RAPA's latest BN2T can carry up to eleven parachutists plus pilot. Here Bob Card, RAPA's energetic secretary at the time, accepts a BN2T – registered G-BOBC in his honour – from Guy Palmer, then Britten-Norman's marketing director.

An Army Air Corps Islander AL Mk1 is seen almost in the centre of this 'stack' of aircraft types flown by the British Army over the years, which includes, from bottom, Chipmunk, Auster, Beaver, Defender, Skeeter, Sioux, Scout, Gazelle and Lynx aircraft. The Britten-Norman aircraft was seen initially as a Beaver replacement.